Excavation & Grading Handbook

By Nick Capachi

Craftsman Book Company
6058 Corte Del Cedro, Box 6500, Carlsbad CA 92008-0992

©1978, 1979, 1980, 1981, 1985 Craftsman Book Company

Library of Congress Cataloging in Publication Data

Capachi, Nick, 1934—
    Excavation & grading handbook.

    Includes index.
    1. Excavation.   2. Road construction.   3. Earthwork. I. Title.
TA730.C28   624'.152   78-3850
ISBN 0-910460-54 -X

Fifth printing 1985

Contents

1

Road Survey Stakes

Most road building begins with a set of plans and a survey of the area where the road is to be built. A survey crew will usually set out point-of-origin stakes called *hubs* which identify a point on the ground. The top of the hub establishes the point from which grades (soil elevations) and distances are computed. Distances are measured from these hubs and other reference stakes. Beside each hub will be an *information stake* which explains in surveyor's code what grades are to be established and the distances to them.

Figure 1-1 illustrates typical markings that might be found on an information stake. Usually these are called *cut* or *fill* stakes depending on the type of excavation required. Both sides, the front and the back of the cut stake are illustrated. Below the stake is a cross section of the existing grades and final road grades that are described on the stake. Note first the stake front in the upper left of Figure 1-1. The

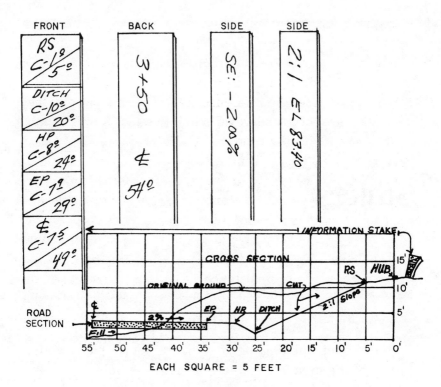

Cut stake reading
Figure 1-1

RS means that the reference stake is the point from which measurements are made. Note the reference stake to the left of the hub. The two horizontal lines mean *and then*, indicating that all the following measurements and elevations are taken from the RS point and not the nearby surveyor's hub. Note particularly that if the double *and then* lines are absent, all measurements and elevations are taken from the hub set by the surveyors rather than the reference stake.

Below *Rs* is *C-1º* written along the diagonal line. The

information above the diagonal line identifies this as a particular point referred to on the road plan and the amount of cut or fill needed to build that point. Below the diagonal line is the distance in feet to the point on the road to be cut or filled. In this case, the reference stake shows a *cut* one foot below the level of the surveyor's hub five feet from the hub.

The ditch cut location is the next information that appears on this stake. It indicates that the ditch is ten feet lower than the RS point and twenty feet out. The grade falls 10 feet over a horizontal distance of 20 feet, thus creating a 2:1 slope. For every foot of cut the grade line moves horizontally 2 feet to arrive at the ditch point by the time vertical 10 feet have been cut. The next reading is the *hinge point* (HP) grade and distance. Note the hinge point on Figure 1-1. It is a lesser cut by 2 feet than the ditch cut. The HP information indicates the grade must come up 2 feet and move out 4 feet. By computing the amount the HP rises from the ditch and the distance it moves towards the center of the road, you can see that it is again a 2:1 slope. Reading down, the next grade and distance is the *edge of pavement* (EP). Notice the cut at EP is .10-foot less than the HP. The reason for this is that the road grade rises 2 percent in the 5 feet from HP to EP. Multiplying 5 feet by 2 percent gives the amount the shoulder rises in that distance ($5.00 \times .02 = .10$).

The next markings give the center line cut. You can see that the cut is again less than the previous cut at EP. Subtracting the 29 feet at EP from the 49 feet to the center line leaves 20 feet. The center line is thus 49 feet from RS and 20 feet from EP. The cut at the center line is .40-foot higher than EP, giving a 2 percent slope from the center line to EP. This is computed by multiplying the 20 feet by 2% ($20.00 \times .02 = .40$).

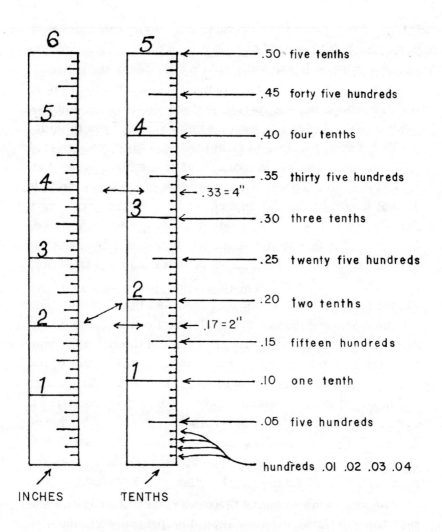

Comparing inches and decimals of a foot
Figure 1-2

Now note the back of the cut stake. It is marked $3+50$, indicating that this is station $3+50$ and is 350 feet from station $0+00$ or the point from which the survey began.

Below the station number is the distance from the surveyor's hub to the center of the road. This includes 5 feet to the RS and 49 feet from the RS to the center line, a total of 54 feet.

Now note the first stake marked *side*. This side of the stake identifies the percentage of slope from the center line to HP. The minus line indicates the center line slopes down to the HP. If it were a plus the center line would be sloping up to the HP. The second stake marked *side* has first the rate the cut slope falls from RS to the ditch. In this case it is 2 feet out for every foot downward. The second group of numbers is the elevation above sea level of the hub.

Notice that all measurements are in feet and tenths of a foot rather than feet and inches. Setting grades requires many additions and subtractions and the use of decimals speeds the work while making errors less likely. Figure 1-2 compares inches with decimals of a foot. The foot is easily divided into both tenths and hundredths.

Figure 1-1 illustrates a cut stake where material must be excavated to reduce the grade to the desired level. Figure 1-3 illustrates a typical fill situation where soil has to be deposited to build up the existing grade. Again, four sides of the stake and the road cross section are shown. The *RS* means that the reference stake (to the right of the hub) is the starting point and the place from which all measurements and grades are established. The cuts and fills given for the RS point will be the existing ground grade at the RS distance indicated. Here, the RS is located 1.8 feet above the hub and three feet from it. The man who sets the grade will have to set the reference stake the horizontal distance indicated from the hub and draw a horizontal line on the stake at the elevation given on the surveyors information stake.

Reading down the stake, the two horizontal lines mean

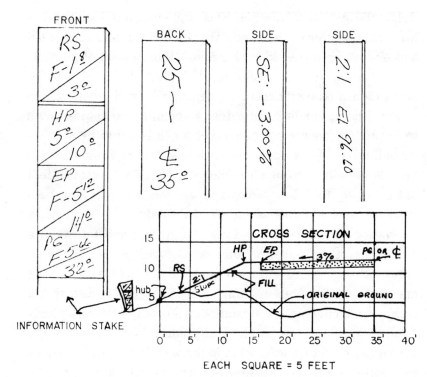

Fill stake reading
Figure 1-3

and then, indicating that the grade setter must measure from the RS point for the next fill and distance given and will not measure or shoot grades from the original hub again. For the *hinge point* (HP), measure 10 feet from the RS stake or lath. At this point a fill of 5 feet must be made to obtain the required grade. The hinge point is the place where the fill slope stops and the road grade begins and is sometimes called the *catch point.* Next reading down the stake is the EP. This is the edge of the pavement and has a F-5.12 fill 14.0 feet from RS.

Below the EP data is PG. This is the projected center line grade. In some cases the surveyors will mark it as the center line and not PG. From the RS measure 32 feet and fill 5.66 feet. This will put the PG or center line 18 feet from the EP and .54 of a foot higher.

The stake marked *back* has a *25* and little worm line standing for +00. Some jobs might have an A, B, C line with one being at the center line. Check the plans to learn what all the designated lines mean. The 25+00 identifies this stake as being 2,500 feet down the line from the point where the measurements started. The point the surveyors start from is most likely marked 0+00, but may not be in all cases.

Next, reading down, notice a C and an L, one over the other. This means *center line.* The number 35 below that means that the center of the road is 35 feet from that point. Look back to the stake marked *front* and notice that when the RS distance of 3 feet is added to the PG distance of 32 feet, the total is 35 feet, the same distance as marked on the back.

The stake labeled *side* is marked SE -3.00%. This is the percent the road bed slopes from the center line to the hinge point. On the far right stake marked *side,* the first reading is 2:1 (two to one). This is the rate the fill slope will rise from RS to HP. Notice that the first stake has a 5 foot fill over a 10 foot distance. This is what the 2:1 indicates. The next item down the stake is EL 96.6. This is the elevation of the hub at the information stake. It is from that hub that all cuts or fills were computed.

What has been described so far in this chapter is more or less standard procedure for indicating elevations and distances. However, surveyors in some counties and cities follow slightly different procedures. Some surveyors provide

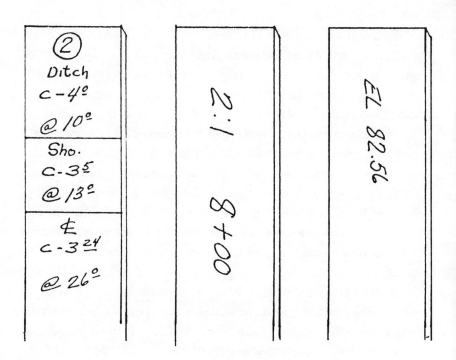

Some county and city stake readings
Figure 1-4

more information on the stakes. The stakes in Figure 1-4 show what might be seen on some county or city stakes. The top of the stake has a 2 with a circle around it. This indicates that the first cut starts 2 feet out. The next markings indicate that the cut is 4 feet at 10 feet. The slope will again be 2:1 because the first two feet are not cut and the cut over the next 8 feet is four feet. Look at Figure 1-4 again. Notice that there is no double *and then* line. This means that you must take all measurements and grade shots from the hub set by the surveyors rather than from an RS, SS, or RP point as on the previous stakes shown. The second group of numbers down gives the top of the shoulder cut. This is the same as the HP on previous stakes. Engineering companies may mark their points differently. Check the plans and the distances on them to determine what points are actually indicated.

The first side stake in Figure 1-4 shows the rate of fall of the cut slope and the station number. The far right stake gives elevation above sea level. In some cases the hub elevation will not be on the stake at all. It may be replaced with the percentage of slope on the road or both may be omitted entirely. The back of the stake in Figure 1-4 will still have the station number but no center line distance because all the front measurements are from the hub and not an RS, SS, or RP point.

Many stakes have very little information but have just the details required to allow you to set the grades. These stakes always have the station number on the back though the percent of slope and hub elevation may be absent. Note Figure 1-5. The left hand stake is what one would expect the surveyor to set for cutting and setting curb grades. From the hub at the base of this information stake move out 5 feet and down 1.50 foot to the top of the curb to set the curb forms. In some cases the surveyors will give the front lip grade or even the flow line grade. If not, it is necessary to determine the distance from the back of the curb to the lip. This information is available in the plans. When setting curb subgrade, determine the thickness of the curb plus any aggregate base if called for under the curb. The thickness of one or both must be added to the cuts and subtracted from the fills to obtain subgrade rather than finished grade.

The center stake in Figure 1-5 is a stake you would expect to find in a subdivision for the first road cut. The front

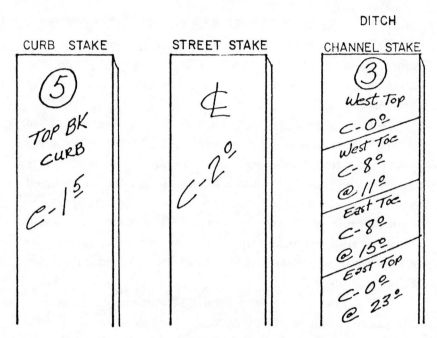

Miscellaneous stakes
Figure 1-5

of the stake indicates the center line of the street and the cut or fill to the finished grade. The grade setter must check the plans for the road width, percentage of slope or crown, and the thickness of the road section. Remember to add the thickness of the road to this cut. On the back of the street stake will be a station number.

The far right stake in Figure 1-5 is a grade stake for a ditch or small channel. The three in a circle is the distance from the hub where the first cut starts. The *west toe grade* indicates the first slope and the bottom of that slope. The *east toe* is the bottom of the slope on the opposite side of the ditch. Both toe cuts are the same so the bottom is flat. The *east top cut* is where the cut will be started on the opposite

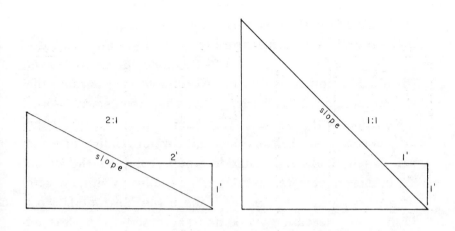

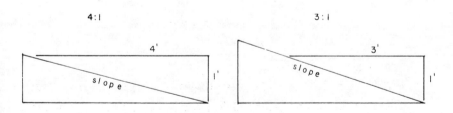

Slopes 1:1 to 1:4
Figure 1-6

side. Subtracting the three foot offset from the 23 foot distance to the east top cut gives the distance across the top of the ditch, 20 feet. Subtract the smaller toe distance from the larger. This gives the width of the ditch bottom, 4 feet. To determine the rate of slope from the top cut to the toe of the channel, subtract the distance given to the top cut from the distance given to toe cut. The three foot offset must be subtracted from the west side distance of 11^{00} feet. This will make the distance 8 feet from top cut to toe on each side. Dividing the C-8^{00} into the 8 foot distance gives an answer of one. This indicates that for every foot cut vertically, the slope moves out 1 foot horizontally. This creates a 1:1 slope.

A stake with only a few markings can give you all the details you need. A little effort and careful reading of plans will usually make it possible for you to complete the work.

In this chapter we have described grades by either a ratio of run to rise or as a percent above the horizontal. Most grades in excavation work are expressed as a ratio of horizontal distance (run) to vertical distance (rise). Figure 1-6 illustrates the four most common slope ratios and should help you visualize most of the slopes you work with in excavation.

2

Plan
Reading

Chapter 1 explained how to read survey stakes. The markings on survey stakes are shorthand or simplified means of conveying the information that is on the excavation plans. As was explained in Chapter 1, you will often have to refer to the plans to fully understand what work must be done. This chapter is intended to make you familiar with the two most common types of excavation plans: plans for subdivisions and plans for highway projects. It is essential that the foreman and grade setter be able to read and understand the plans. It is also an asset to your company to have some operators and laborers who can read plans. Any time the surveyors use an unfamiliar abbreviation or notation on a stake, the foreman or grade setter must turn to the plans and see what is intended.

Figure 2-1 is the first sheet of a set of plans. It shows a typical street cross section plan that you might have on a road excavation job. Notice that every distance is given from

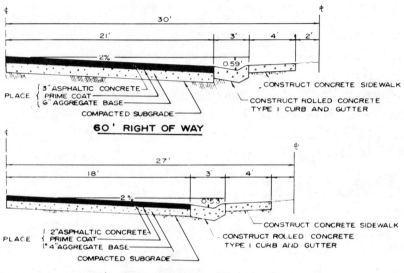

3" ASPHALTIC CONCRETE
PLACE { PRIME COAT
6" AGGREGATE BASE
COMPACTED SUBGRADE
CONSTRUCT CONCRETE SIDEWALK
CONSTRUCT ROLLED CONCRETE TYPE I CURB AND GUTTER

60' RIGHT OF WAY

2" ASPHALTIC CONCRETE
PLACE { PRIME COAT
4" AGGREGATE BASE
COMPACTED SUBGRADE
CONSTRUCT CONCRETE SIDEWALK
CONSTRUCT ROLLED CONCRETE TYPE I CURB AND GUTTER

54' RIGHT OF WAY

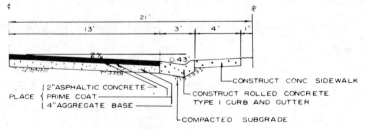

2" ASPHALTIC CONCRETE
PLACE { PRIME COAT
4" AGGREGATE BASE
CONSTRUCT CONC SIDEWALK
CONSTRUCT ROLLED CONCRETE TYPE I CURB AND GUTTER
COMPACTED SUBGRADE

42' RIGHT OF WAY

SHEET INDEX

Typical cross section
Figure 2-1

the centerline: to the front of the curb, to the back of curb, to the back of the walk, to the property line. The number at the top center of each street cross section is the distance from the centerline to the property line. The lower group of figures indicates the distance from the centerline to the lip of the curb, the width of the curb, the width of the sidewalk, and the distance from the back of the walk to the property line. Refer to the top street cross section. Above the street surface and curb the figures 2% and 0.59 can be seen. The figure 2% is the slope of the street surface from the centerline to the lip of the curb. The figure 0.59 indicates that the flowline of curb and gutter is fifty-nine hundredths (0.59) below the centerline of the street.

Below the cross section drawing the type of paving material and the thickness required is indicated. The actual type of asphaltic concrete, prime coat, aggregate base, curb and gutter and sidewalk required will be described in the job specifications. In some cases the job specifications will refer to a master set of specifications. The percentage of compaction required will also be spelled out. in the specifications.

The right-of-way distances (*60', 54'* and *42'*) are the measurements from property line to property line.

The cross section indicates a "type 1" curb is required. The dimensions of this curb and walk will be explained in some county, city, or state standard specifications or drawings.

Below the cross section drawings is a sheet index. This index shows the title of each drawing in the complete set of plans and gives the page number for each drawing. Every large set of plans has an index. Every set of plans will have a typical cross section drawing, whether it is a subdivision street with curbs or a highway with ditches at each edge. This plan is usually referred to as the "typical". The typical drawing usually shows only half of each street. That's all you need when both halves are exactly the same.

The grading plan in Figure 2-2 may at first glance seem to be very complicated. Actually it is fairly simple to understand. This is a portion of grading plan for a subdivision and supplies all the information needed to grade the lots. Each lot is numbered and has a finished lot pad elevation indicated. The lot numbers are the smaller numbers from 399 to 470.

Notice lot number 444, slightly to the left of center in Figure 2-2. The elevation of the building pad is 161^5. This indicates that the pad is one hundred sixty-one feet and five tenths above sea level. The larger the elevation number, the higher the pad. Lots 443 to 445 are level. Lots 446 to 448 get progressively lower. Lot 448 is six feet one tenth (6^{10}) lower than lot 445. To find the difference in lot elevations, subtract the smallest elevation from the largest elevation.

Lot 444 has x's marked in each of four corners. Notice also two dots in front of the two front lot x's. The surveyors will set hubs and stakes where the two dots are marked. These hubs will be 5 feet from the x's which mark the front lot pad corners. Each stake will have a 5 foot offset marked on it and indicate the amount of cut or fill required for the spot where the x is marked on the plan. The back lot stakes are usually set at the rear lot corners where the back x's are marked. No offset will be marked on them.

The elongated *y*'s at the front and back of lot 444 indicate that the ground slopes down from the lot pad to the back of the walk in front and to the lower lots at the back. Notice the same elongated *y*'s at the sides of each lot that is above or

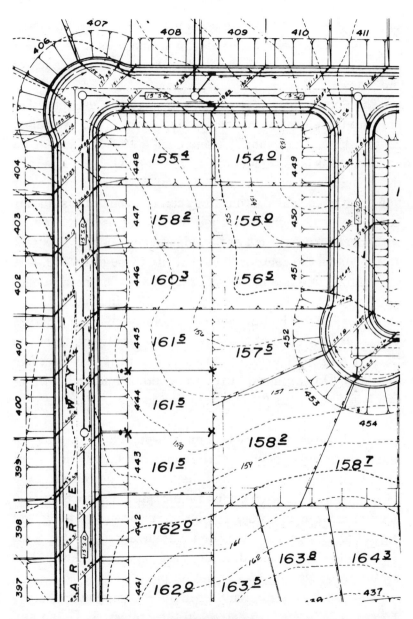

Grading plan
Figure 2-2

below the adjoining lot. This helps you visualize the slope.

There are several small elevation numbers in the street area. These elevations indicate the level at the top back of the sidewalk.

The dotted lines that wander through the grading plan indicate the elevation of the existing ground and are called contour lines. Occasionally along each dotted line you see a number. This is the elevation above sea level. Each line is at the given elevation no matter how many turns it makes or how far it goes. The contour lines can tell you approximately how much cut or fill is needed on any lot or in the street without checking the cut or fill indicated on each lot or street stake. Notice the contour lines that run through lot 443. The top contour line has an elevation of 158 ^{0}. The bottom contour line, if followed to the left to lot 453, indicates it is at elevation 159 ^{0}. The finished lot pad elevation is indicated as 161^{5}. Subtract each of the two contour elevations from the finished lot pad elevation. You see that a fill of three feet five tenths (3^{50}) is required at the top contour line and a fill of two feet five tenths (2^{50}) is required at the bottom contour line.

Check the contour lines running through lot 453. You will notice that a fill is needed at the top of the lot pad and a cut must be made at the bottom of the lot pad. The center of the lot is near the finished lot pad elevation before excavation begins. A foreman studying the contour lines could color all the cut areas one color and the fill areas another color. This would help you see the fill and cut areas easily and would be useful in developing an excavation plan.

There will be a plan and profile sheet similar to Figure 2-3 (A) and (B) for every street in the project. The plan and profile sheet adds information not given on the grading plan or the street cross section drawings. The Figure labeled 2-3 (A) is the profile of a 500 foot section of Zenith Drive and includes a plan for the sewer and drainage lines. Profiles are drawn on graph paper. In this case, each square represents one foot vertically and 10 feet horizontally.

Going up the right side of the graph paper are figures from 140 to 170. These figures represent the elevation above sea level. At the bottom of the graph are six figures from 29 to 34. These are the one hundred foot stations indicated along the centerline of the plan drawing, Figure 2-3 (B).

At the bottom right corner of the graph paper appear the centerline symbol and the words "finished grades". This means that all the figures above along the right margin of the page are finished road centerline elevations. At the left of the graph paper written horizontally are the precise station numbers the right elevations relate to. Notice the words "natural ground", "finished grade" and "gutter grade" near the right center of Figure 2-3 (A). The first line says "natural ground" and has a line and arrow pointing to an irregular horizontal line. This indicates the original ground elevation before any excavation has been done. The second line has the center line symbol and the words "finished grade". The line and arrow point to a heavy horizontal line. This indicates the elevation of the finished centerline of the street surface.

Notice that the slope of the centerline is indicated in three spots where the slope changes. These are -0.87%, -2.30% and -.73%. Along the centerline finished grade line there are four small circles. Follow down the vertical graph lines that split those circles until they reach the black triangles. Between each set of black triangles a distance is given (100 feet and 80 feet) and the words "vertical curve"

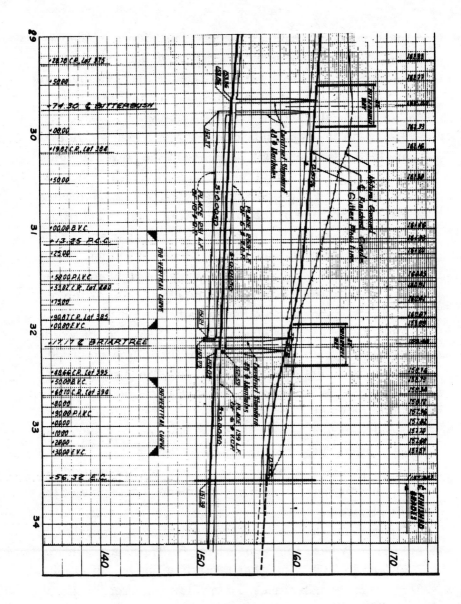

Profile
Figure 2-3 (A)

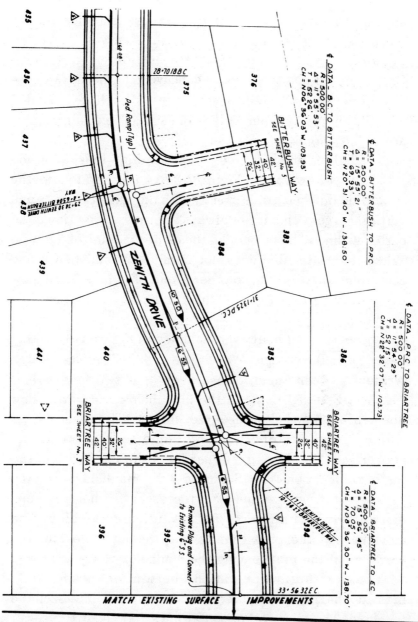

Plan
Figure 2-3 (B)

appear. This indicates that between those triangles and for the distances given the road centerline curves. Notice in Figure 2-3 (A) that the curves are in different directions. One is convex and the other is concave. The portions of the centerline finished grade line that are straight are labeled with the appropriate percentage of slope. The straight areas are called *tangents*.

The lighter line just below the centerline finished grade is the gutter flow line. In Figure 2-3 (A), this line follows the centerline finished grade exactly with the same tangents and vertical curves. This is not always the case. Many times the curb flow line will not parallel the centerline finished grade. When the gutter line does not parallel the finished grade centerline, there will be another group of figures just below the centerline finish grade which indicate gutter flow line elevations.

Locate the four heavy vertical lines on Figure 2-3 (A) that have "42' Bitterbush Way" and "42' Briartree Way" written between them. These show that two side streets intersect Zenith Drive. All the lines below the gutter flow line indicate pipe below the street surface. The vertical lines which are close together and come to a point at the centerline finished grade are manholes. Notice the instructions "Construct Standard 48" manholes". At the bottom of the vertical lines that represent manholes are one or two sets of numbers. These indicate the flow line elevations of the pipe coming into the manhole and the elevation of the pipe leaving the manhole.

The horizontal lines at the bottom of the manhole indicate the pipe and direction of flow that intersect the manhole. The top two horizontal lines represent the sewer line and the wider bottom two lines show the drain line.

Notice the writing with arrows pointing to the sewer and drain lines. "Place 253 LF of 6" V.C.P." means that between the two manholes there will be 253 feet of 6 inch vitrified clay pipe. The writing directly below says "Place 231 LF of 10" C.P." This means that between the two manholes there will be 231 feet of 10 inch concrete pipe. The percentage of slope along each pipe section is given. In this case the slope is indicated as "S = 0.0050". With the rate of slope given for pipe, street, and flow line of the gutter, the amount of slope per foot can be computed. This is done by multiplying the percentage of slope by the total length in feet and then dividing the total length into the answer. In this case: 0.0050 x 253' = 1.265' ÷ 253' = 0.005. The sewer line would drop 1.265 feet in 253 feet and 0.005 foot in each foot of length.

Look now to Figure 2-3 (B). This is the plan drawing of the profile in Figure 2-3 (A). Find the small station numbers from 30 to 33 along the centerline. Notice also the dotted and solid lines running perpendicular through the centerline with a small circle where the two lines meet. At the far left you will see the writing 28 + 70.18 B.C. to the left of the line. The B.C. is the abbreviation for begin curve. At the far right side and at the top of the street there is written 33 + 56.32 E.C. E.C. is the abbreviation for end of curve. A station number is written on a line between the B.C. and E.C.: 31 + 13.25 P.C.C. This indicates Point of Curve Change. There are five circles with station numbers at each. These are distances the surveyors must know to stake the curve. Notice the four groups of centerline data written across the top of the sheet. The data supplies the surveyors with the distances and transit readings needed to turn the correct angles to stake the centerline curve. No radius point is set for a curve of this type. There will be radius points set at each corner where the two dotted lines on each corner meet.

The plan drawing shows the lot location and numbers but does not give lot elevations as found in the grading plan, Figure 2-2.

The plan drawing of Zenith Drive shows a sewer main and a storm drain line in the street section. There are three pointed box symbols written in the street area that indicate the size of pipe and the direction of flow. The symbol at the left marked *10" S.D.* means that the pipe has a 10 inch inside diameter and is a storm drain. The other two pipe symbols are marked *6" S.S.*, meaning a 6 inch inside diameter sanitary sewer line. The black triangle tips of the symbols show the direction the line is flowing.

Sewer service lines off the main line are indicated with a small triangle symbol with an *S* in it. The circles drawn in the main lines represent manholes. Notice the 10" S.D. ends at the intersection of Zenith Drive and Briartree Way. From the manhole at that intersection four lines are shown radiating out from the manhole. These end at four small black rectangles that stand for gutter drain inlets.

The water system is not shown on this plan. Usually the water system and the electrical street lighting layout are on a separate sheet.

The plan drawing for a highway is similar to the plan drawing of a subdivision. See Figure 2-4 (A) and (B). After looking at Figure 2-4, turn to Figure 2-3 and examine the subdivision plan drawing and profile. Notice that they are very similar. The highway profile, Figure 2-4 (B), is very simply drawn. Notice the elevations from 16 to 18 on the top edge and the station numbers from 195 + 00 to 200 + 00 along the left edge. Each rectangular segment of the graph paper used for this profile represents .20 (two tenths of a foot) vertically and 50 feet horizontally. Check the rectangular segments with the station numbers and elevations given and you will see that this is so. The *N* at the left of Figure 2-4 (*A*) show north.

The irregular horizontal line through the middle of the graph paper is called the *existing h line*. This is a reference line used in drawing the plan and profile. The architect in this case designated the centerline of the road as the "H line" rather than the "centerline". This profile does not show the elevation of finished H line. Usually the highway profile drawing has more details and would be more like the profile drawing for the subdivision, Figure 2-3 (A).

A finished H line is not shown because except for the asphalt overlay, all the work is to be done adjacent to the existing pavement. Notice Figure 2-5 (cross section). It shows all the ditch and road section excavating adjacent to both sides of the existing road. The only time the existing H line shown in Figure 2-4 (B) will change is when the shoulder widening is complete and the final asphalt pass is made over the existing road and H line. Look at the cross sections in Figure 2-5. Notice the existing H line elevations written vertically at each station. The second elevation written on an angle and underlined is the finished H line elevation. Notice the largest grade change at any station between existing H line and finished H line is .20 foot. This profile drawing also does not show any underground pipe.

Notice the left side of the highway profile says "ditch Lt" and ditch Rt". *Rt* means right and *Lt* means left. The ditch elevations for both sides of the road are given at 50 foot intervals. The right side of the road is the side that would be

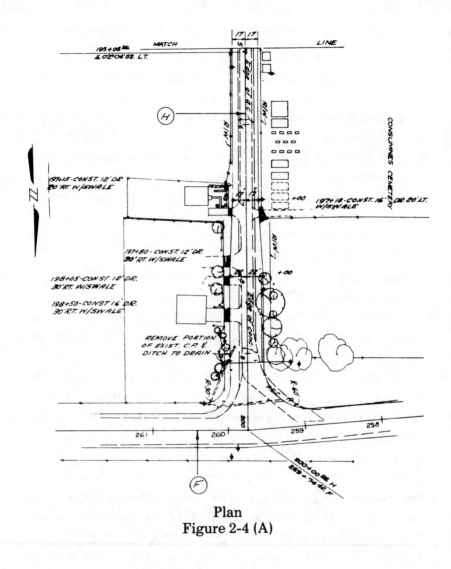

Plan
Figure 2-4 (A)

on your right if you stood with the smallest station numbers
to your back and looked down the road toward the larger
station numbers. In Figure 2-4 (A) the right side of the road
would be the west side of the road and the left side would be
the east side.

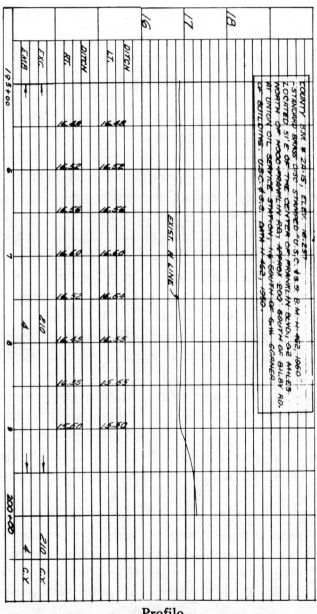

Profile
Figure 2-4 (B)

At the bottom left of the profile are two abbreviations: *Exc* and *Emb*. These stand for excavation and embankment. Look across the sheet to the figures *210 c.y.* and *4 c.y.* written at the lower right. This means that there are an estimated 210 cubic yards of excavation and four cubic yards of embankment in this section of roadway. Most highway profile drawings have the estimated yardage figured to help the foreman and estimator. A glance at each page of the profile will show you where the major cuts and fills will be made. A foreman looking at the figures in Figure 2-4 (B) would see that there will be 206 cubic yards of excess dirt to haul off or move to an embankment further down the road.

Read the information in the rectangular box at the top of Figure 2-4 (B). This explains how to locate the bench mark which the surveyors used to establish the grade for the project. The elevation given, 16.237, was rounded off to 16.24 or 16 feet and 24 hundredths above sea level. Every job is staked with a point of known elevation called a bench mark.

Refer again to Figure 2-4 (A). The station numbers on the H line correspond with the station numbers in the profile drawing. The width of each half of the street is shown: 17 feet at stations 5 and 7 and 22 feet at station 8. You should understand that station 8 is more fully described as station 198+00.

Notice that Figure 2-4 (A) shows the location of all the trees, fences, houses, and two sections of pipe at the intersection. At station 9 there is a line crossing the road with a small *T* inset. This indicates that an underground telephone line crosses at that point. The width of the existing road is pointed out in two places with the words *Edge of A.C.* (asphaltic concrete) and *Edge of Conc.* (concrete). The

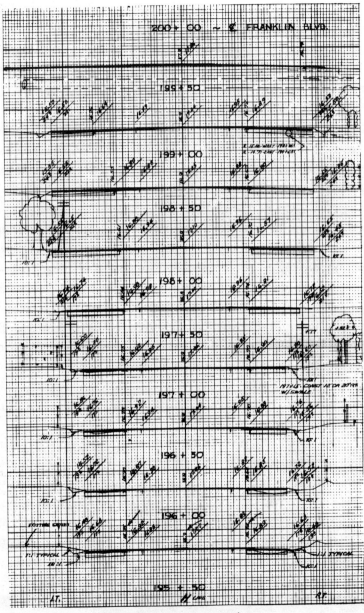

Highway cross section
Figure 2-5

writing at the left and right of Figure 2-4 (A) gives the
details of where the driveways must be built. The driveway
instructions at the right of the sheet give the station number
for the center of the driveway, 197 + 18. It then indicates that
a 16 foot wide driveway must be built at station 197 + 18. The
driveway is to extend from the edge of the new road to 20
feet left of the H line. A swale is a shallow dip in the
driveway to let the ditch water pass through. A swale is used
rather than placing a pipe through the driveway to let the
water pass under.

Every plan and profile drawing such as Figure 2-4 (A)
and (B) must have a cross section drawing so that you have
the information you need to build a highway. Typical cross
section drawings are shown in Figure 2-5. Notice that the
station numbers in Figure 2-5 are the same as the station
numbers in Figure 2-4. Figure 2-5 is a cross section view of
the highway in Figure 2-4 at every 50 foot station. The
smallest station numbers are at the bottom on the cross
section drawing. Visualize this as being as though you were
standing in the road with the small stations at your back.
This puts the *Rt* ditch on the right side of the sheet and the
Lt ditch on the left side of the sheet.

The cross section drawing is on graph paper. The heavy
black lines are at 10 foot intervals. Each small square
represents a 6 inch square. Several numbers are written
diagonally above each cross section. These are finished road
and ditch elevations. Below the two diagonal lines at left and
right side of each cross section there is a number. This is the
distance in feet to the edge of the pavement and the
centerline of the ditch. The distance on the right and left
from station 195 + 50 to 197 + 00 is 17 feet to the edge of the
pavement and 19 feet to the ditch. From station 197 + 00 to

199+50 the ditch and edge of pavement widen or move
further from the centerline and reach 22 feet to the edge of
the pavement and 24 feet to the ditch at station 199+00.
These distances are measured from the centerline or as in
this case, the H line.

Numbers are written vertically and pointed out by arrows
above station number 195+50. These are the elevations of
the existing road surface at the centerline and at each edge
of the pavement. Study the cross section drawing. Notice
that full structure sections are to be built on each side of the
existing road. Follow the dark irregular line running
horizontally through the drawing. This line is the existing
ground elevation before the ditch and road sections are
built. The existing ground line is pointed out in the drawing
at station 195+50. The rate of slope for the front and back of
the ditch is indicated as *1:1 typical* and *10:1* respectively at
station 195+50.

A note at the right side of the drawing of station 197+00
indicates that a driveway is to be built at station 197+15.
Twenty feet right of centerline is to be the limit of
construction. Above at station 199+00 on the right side is
information for removing drain pipe. The flow line elevations and station numbers of each end of the pipe are given.
Subtracting the small station number from the larger number gives you the length of pipe to be left, in this case 773
feet.

At the right and left of each drawing several trees and
fences are drawn in. This is done to insure that the
contractor leaves them in place and to show the clearance
between these objects and the ditch slope. In the ditch areas
on each side of the drawing at station 197+00 a power pole
is shown. At station 198+00 another power pole is indicated

in the new section to be built. That pole must be moved before or during construction. In some cases you will have to work around a pole.

Along with the plan drawing and cross section drawing, Figures 2-4 and 2-5, the highway plans will include a typical road section sheet similar to Figure 2-1. Always study these plans carefully before beginning a project.

You can make difficult plans easier to read if you color code them. For example, all the new widening areas could be colored green, the overlay areas red, and the pipe runs yellow. This would help you distinguish each excavation area more easily.

Occasionally you find a symbol or notation that doesn't make sense to you or that you have not seen before. Even experienced excavation contractors may come across something on a plan that is not clear. In every case the surveyor or engineer that developed the plans should be able to answer your question.

3

Grade Setting

The job of setting grades is one of the most critical jobs on a construction project. The grade setter transfers information on the plans and surveyor's stakes to stakes that his own equipment operators can read and follow. The grade setter must be both fast and accurate: fast to stay ahead of the equipment and accurate because any error can be very costly. In many cases the foreman will not have time to check the grade setter's work and any error may not be discovered until days later when it becomes very obvious to the foreman and grade setter alike.

A good grade setter must be able to perform rapid mental mathematics. His entire day is spent adding, subtracting, and multiplying as he transfers grades and distances off the original surveyor's stakes and sets his own stakes and grades. He must set his stakes at every strategic grade change. They must not be in the way of the equipment yet must be in a spot where the operator can see them. A grade setter must stake the tops of slopes to be cut, the slopes themselves as they are trimmed to grade, and the toe of the slope once it is reached. He must set stakes at ditch bottoms, road shoulders, road centerlines, and so on.

In most cases the grade setter will not drive a hub into the ground because hubs are usually used only for trimming to a close tolerance. Instead he will drive a lath from 1 to 4 feet long. On the lath he will usually have markings similar to 1, 2 or 3 in Figure 3-1 (A).

(1) The *crows foot* or *grade lath* is the most common way for the grade setter to show the grades to the operators. The circle through the tail of the arrow always indicates that the point of the arrow is drawn to the finish grade. When the operator sees it he knows the finish grade is at that point.

(2) The crows foot or grade lath in the middle of Figure 3-1(A) does not have the circle drawn through the tail of the arrow. This indicates to the operator that grade has not been reached yet. By looking to the top of the lath he will see that five tenths of a foot (6'') must still be excavated below the line marked across the lath.

(3) This crows foot or lath is similar to number 2. However, this lath would be on a fill slope. It gives both the fill still needed at the lath (up to the horizontal line) and the fill needed 10 feet from the lath (five feet). The 2:1 slope was computed by dividing the 5 feet left to fill into the 10 foot distance from the stake.

Setting up a line of laths and marking them is usually called *setting crows feet*. This method is faster than setting hubs because the top of the stake is not set on grade. In most jobs the equipment will eventually be directed to make a pass right over the lath line. Then the lath will have to be re-set. Where an extremely accurate grade must be cut, the

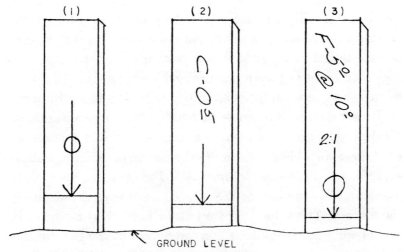

**Crows feet
(Set by the grade setter)
Figure 3-1 (A)**

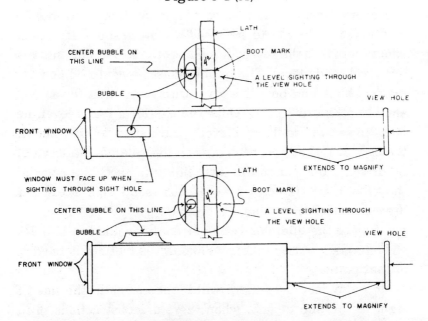

**Two types of eye levels
Figure 3-1 (B)**

grade setter will replace his laths with hubs once the grade is nearly established. The hub marks a point on the ground like a lath but also establishes a precise elevation at the top edge. After setting the hubs he should drive a short lath next to each hub so it can be seen by the men on the equipment.

To set a crows foot grade or a hub, the grade setter must establish the ground elevation at a given point in relation to the surveyor's hub. Assume the surveyor's information stake reads *C-350 centerline 350*. The grade setter must measure from the hub or R.S. stake, whichever is indicated, the 35 feet. There he drives a lath or hub. Next he sets his ruler (small numbers down) on the hub or at the base of the lath. If a hub is used it should be only partly driven in the ground at this time. The grade setter will then add in his head the road section thickness to the cut computed by the surveyors. Assume the road section is .50 foot. Add the .50 to the C-3^{50} + .50 to get C-4^{00}. The grade setter then shoots across to the surveyor's hub or the R.S. point until his eye level reads level. The bubble must be centered and his eye level line and boot line level with each other for a level shot. See Figure 3-1 (B). Once the eye level reads level, he looks to see where the eye level is in relation to his ruler. If the eye level is at 3 feet 8 tenths on the ruler, the hub must be driven .20 lower so the level line intersects the ruler at the 4 foot mark. If a crows foot lath is used, the grade setter draws a horizontal line at ground level with an arrow pointing to the line. Above the arrow he would mark C-0^{20} so the equipment operators will know .20 foot cut is needed at that point.

In many instances there will be obstacles in the line of sight to the hub or R.S. when there are obstructions. The grade setter sets up *boots* before any work begins. To set a boot, the grade setter drives a 4 foot lath behind the hub the surveyors have set. Then he rounds the surveyor's cuts off to an even foot. The C-3^{50} the surveyors set would be rounded up to 4 feet. The grade setter measures from the hub up .50 foot and draws a horizontal line on the 4 foot lath .50 foot above the hub. If he feels he needs to measure up the lath further to clear any obstacles in his line of sight, he may raise the boot another foot or more. Assume that .50 foot is enough to clear all obstacles. Then from the horizontal line marked on the 4 foot lath there will be a cut of 4 feet to the finished grade. The grade setter marks all the lath from the surveyor's hub up the distance needed to round the cuts off to 4 feet. For example: C-3^{70} + .30 = 4 foot. C-2^{60} + 1.^{40} = 4 foot. C-3^{10} + .^{90} = 4 foot. If a fill rather than a cut is marked, measure up the amount of fill marked by the surveyors + 4 feet. For example: F-0^{10} measure up 4^{10} = 4 foot boot. F-0^{40} measure up 4^{40} = 4 foot boot. F-0^{05} measure up 4^{05} = 4 foot boot. Doing this at every hub makes the grade setter's job easier and faster. Adding the .50 road section thickness to the ruler reading at each station will give a 4^{50} cut at every station that is set.

On many occasions the grade setter cannot set up laths because there is no room for them. He must then check the grade after each scraper or grader pass and let the operator know how much more cut or fill is needed.

Refer to Figure 3-2, "Staking for a typical cut and fill section". The only hubs set are those set by the surveyors. From these the cuts, fills, and distances were computed. The surveyors first had to establish the center line and stake it as shown. From these stakes the existing ground elevation at the edge of the shoulder could be established. The surveyors could then determine the cut or fill needed at that

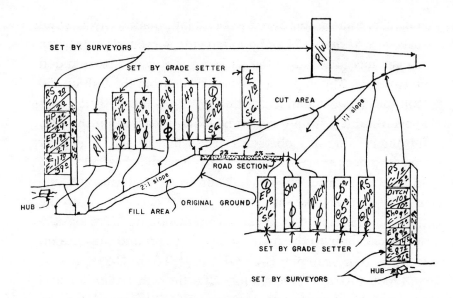

Staking for typical cut and fill section
Figure 3-2

point from the elevations shown on the job plans. The amount of cut or fill must be known so the surveyors can compute the distance needed for slopes and ditches from the edge of the shoulder to the point where the hub must be set on each side of center line.

Assume that the surveyors have computed the distances needed and have driven the hubs required on each side of the center line at each key station. Now they shoot the elevation at every hub. Once the elevation of every hub has been established the cuts, fills, and distances needed to build the road from that hub back to center line will be computed and marked on the information stake next to the hub. The surveyors at this point have supplied the contractor with all the information needed to build the road.

Look again at Figure 3-2. Except for the *R/W*

(right-of-way) stakes, all stakes have been set by the grade setter and all are crows feet. No hubs have been set by the grade setter. Notice the two crows feet marked *Sho* (shoulder) and *H.P.* (hinge point). Some grade setters set hubs at those points if the grader operator doing the trimming needs or requests them for more control and a closer cut. The shoulder and hinge point grade will be fine trimmed after the road has been paved and should be left a little high at this time; two tenths of a foot (.20) would be about right.

Once the dotted area in Figure 3-2 labeled *road section* has been cut out, the crows feet at the edge of the pavement and at the centerline will be replaced with hubs so the subgrade can be compacted and fine trimming can be done accurately.

The grade setter should give some thought to the best place to locate the stakes. Also, timing can be as important as the actual location. No matter where or when the stakes are set, keep the stakes or hubs set on firm ground or aggregate. For example, suppose hubs are needed in an area where fill has been brought up to the grade marked on laths. The grade setter must be sure to pull the stakes or offset them so the area where the stakes had been can be compacted before setting the hubs. In a deep fill area the lath must be offset periodically so the area where the grade laths are placed can be filled and compacted and new lath set back with grades for the remaining fill. Never set stakes so close to the area being filled that the edge of a fill cannot be compacted without covering the stakes. All fill hinge points must be over-filled so that when trimmed the hinge point will be well compacted. If the shoulder is to be rocked or paved and the subgrade at a hinge point was not

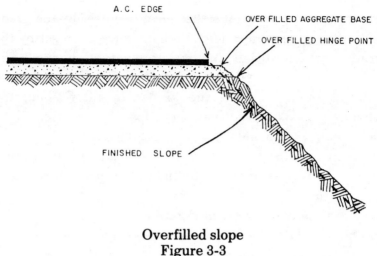

A.C. EDGE

OVER FILLED AGGREGATE BASE

OVER FILLED HINGE POINT

FINISHED SLOPE

Overfilled slope
Figure 3-3

over-filled slightly, the rock or pavement will be lost over the edge of the slope. See Figure 3-3.

If a dike is to be placed on the edge of the pavement, the grade setter or foreman must check the edge of pavement (EP) distance on the surveyor's stake. The surveyor's stake may give the distance to the front of the dike or back of the dike. There must be 3 or 4 inches of asphalt concrete behind the dike so the dike machine can function properly. The grade setter should adjust his stakes to give enough of an offset to allow for the extra 3 or 4 inches needed to place the dike.

The grade setter must be very careful when setting lines and grades on all edges. Top of slopes, toe of slopes, hinge points, choker lines, and ditch lines must be set precisely.

The grade setter should never assume the correct cut or fill was made by the equipment operator. After the operator has made the cut or fill, the grade setter should re-check the grade. Both the grade setter and the foreman should watch closely as fills and cuts are made. In many instances, operators overlook an area where a fill or cut must be made. This will usually happen in areas that are more difficult to fill or cut. The grade setter should be sure these areas are cut or filled before the equipment moves to a new area. Moving the equipment back later is wasteful. It is up to the grade setter to see that no small areas or edges are left half finished. He must not hesitate to complain to the foreman if the equipment operators are careless about running over stakes. The grade setter is busy enough without having to replace stakes that were carelessly run into the ground.

The grade setter needs the right equipment to be efficient at his job. He should have a pouch to carry a 50-foot tape, a marking pencil or crayon, ribbon or red spray paint, and a ruler. On his belt he should have a hatchet and an eye level. The grade setter also needs a sack or carrying rack to carry stakes and hubs. If working in hard ground, a steel pin may be needed to drive a hole for the wood hubs. He needs some fluorescent red paint to spray the hub tops and lath tops so they can be easily seen. Spray paint is much faster than tying colored ribbon on the laths or marking blue crayon on hubs. Spray painted hubs are much easier to find after the grader operator has passed over them.

Clothes pins can be used to indicate cuts and fills. This is usually the most efficient method for three reasons. First, it eliminates one man, the guinea hopper, who would otherwise be required to point out the hubs to the grader operator. Second, it saves much of the time normally required to set grade hubs. Finally, it saves the cost of extra hubs and laths. Clothes pin grade setting is going to be the quickest and least expensive method of setting grades any time there are curbs on both sides of the road.

This street has only clothes pin grade stakes with no center
line hubs. The clothes pin indicates that there is a slight cut
remaining at the center line. When the center line is on
grade, the clothes pin will be set on the ribbon. Figure 3-4
(B) shows a close-up using crayon rather than ribbon.
Figure 3-4 (A)

The grade operator must trim one side along the curb to
grade. The grade setter then drives a 4 foot lath against the
edge of the curb at 50 foot intervals. After driving the lath,
the grade setter marks a horizontal line on the lath,

Clothes pin grade lines
Figure 3-4 (B)

measuring up 3 feet above the top lip of the curb. A colored survey ribbon or crayon works well for a horizontal mark. See Figure 3-4 (A) and (B).

If there are no curbs, set the 4-foot lath one foot behind the point to be trimmed at the edge of the road and opposite each grade station. For example, if the surveyor's information stake gives a distance of 20 feet to the edge of the pavement, measure out 19 feet and drive the 4-foot lath.

Next check the elevation given for the edge of the pavement. Mark the horizontal line on the 4-foot lath 3 feet above the level of the finish edge of the pavement. Follow the same steps at every grade station. Then use the procedure explained in the previous paragraph for setting the grades.

Once all the laths are placed and the horizontal lines or ribbons are set, the grade can be checked. If the road has a crown in it, the grade setter must subtract the amount of crown from the 3 foot horizontal line and add the thickness of the road section. This gives the ruler reading desired at the center line. For example, assume that there is a .30 crown. Subtracting it from the 3 foot mark on the lath equals 2.70 feet, adding .50 for the thickness of the road equals 3.20 feet. The 3.20 is the height the grade setter should sight with his eye level at each station to be on grade. The grade setter measures out from the lath to the center line. He then sights back with his eye level along a level line to the horizontal line on the lath. This is done by standing the ruler up with the low numbered end down. Slide the eye level along the ruler until it sights level. Once it sights level, read the numbers on the ruler at that point.

Let's assume that at that point the ruler reading is 3 feet. To be on grade, the ruler should read 3 (feet) and .20 (two tenths). This indicates .20 (two tenths) must be cut from center line before the correct grade is reached. The grade setter will now clip a clothes pin on the 4 foot lath .20 (two tenths) above the horizontal mark. Then the grade setter moves to the next lath and follows the same procedure. This time he reads 3 (feet) .70 (seven tenths) on his ruler. He then would clip a clothes pin on the second lath .50 (five tenths) below the horizontal mark. This means a .50 (five tenths) fill is needed. The grade setter will continue shooting each lath, clipping a clothes pin above the horizontal line for cuts and under the horizontal line for fills. If the ruler reading is less than 3.20, a cut must be made. A ruler reading over 3.20 means that a fill is needed.

The grader operator can see the lath and knows whether a cut or fill is needed by where the clothes pin is in relation to the horizontal mark on the lath. The grade setter need not stay with the grader operator. When the grade setter finishes marking the last station, he should double back and check the work the grader operator has completed. The grade setter will adjust the clothes pin after re-shooting the grade. Once the grader operator finishes the first pass, a second pass can be made starting from the beginning. The grade setter will have adjusted the clothes pins, placing them the exact distance above or below the horizontal line as needed for a cut or fill.

When the center line is on grade, the clothes pin is placed on the horizontal line. When the grader operator sees the clothes pin clipped to the horizontal mark, he knows that no further work is needed at that station. Survey ribbon tied around the lath in place of horizontal crayon marks works well because it can be pulled off when finished. The lath can then be used again.

If the road is wide and the grader operator has trouble determining exactly where the center line is because he has no center line hubs to follow, the grade setter should mark the center line with spray paint each time he takes a grade check. This method of checking grades can also be used for roads where no curbs are present.

Center line and shoulder grade can both be set on the same lath by using two clothes pins painted different colors. One color can be used for the center line grades and another color for the shoulder grade.

There are many types of excavation work where the grade is set best with a string line. This would include laying pipe, setting forms or preparing for a curb machine. The string line should be set with great care and stretched as tight as possible. A good nylon line is best for this type of work. In most cases pipe laying requires a string line. Before setting the line, determine how much of an offset is needed so the string will not interfere with the equipment to be used. If a trencher is used, the string line must be on the operator's side—the side opposite where the spoil is to be thrown. If a backhoe is used, be sure to keep the string line back far enough to clear the equipment outriggers.

When setting string line, keep three stations set up along the line so you can sight down the line to correct errors or movement which might occur during setting or excavation. A faulty station grade will be easily detected this way. When you sight down the string line, each station should blend with the next with no sudden rise or dip from one station to the next. If you detect a sudden grade change, check the measurements. Check the rate of slope shown on the plans if you can't find any error in your work. If the plans do not show a sudden grade change, the surveyors have made the error and it should be called to their attention.

When waiting for the surveyors would hold up production, correct the problem by the following method. Set up three more stations with the string line and sight through the obviously faulty station, lowering or raising the string until it flows smoothly into the correct grade. Usually the surveyors will have made an error at only one station and this will solve the problem without holding up the trenching operation. Always be sure the string line is straight. Look for any slight variations. The trencher, hoe, or automatic machines will use the line for direction as well as elevation.

If the steel pins used to hold the string line are offset to one side from the grade hub, use a carpenter's level and straight edge to transfer your grade from the hub to the grade pin. An eye level may be used, but this is not good practice where a close tolerance is required. If the distance is too far for a straight edge, a surveyor's level is suggested.

Assume now that you are ready to transfer the surveyors' grades to the grade stakes. The surveyors have set stakes for 500 feet of 30 inch drain pipe. The distance the surveyor's stakes are offset from the actual trench location and the side of the offset should have been requested by the contractor. The width of the trench and direction the spoil will be thrown determines these two factors. Let's assume the surveyors have set all their grade hubs on a 10 foot offset. A trencher is to be used for the excavating. The distance the string is offset from the surveyors' hubs is determined by the type of trencher. Measure from the center of the bucket line to the grade indicator arm. This arm extends out along the bucket line and can be adjusted in height. A metal wand projects horizontally from the end of the grade indicator arm. Once the required depth is reached, the grade indicator arm is lowered until the wand is touching the string. It will be kept in that position by the operator by raising or lowering the bucket line. A rod extending from the engine end of the trencher is used to keep the direction of travel straight.

If the measurement from the center of the bucket line to the end of the wand is eight feet, the string line should be set two feet from the surveyor's hub. With the hubs at a 10 foot offset, this will put the string line 8 feet from the center of the trench.

For a job like this there should be no less than four grade

pins set at 50 foot intervals along 150 feet of trench at any time. Once the string line pins are up, the grade setter must transfer the grade given by the surveyors to the grade pins at an even foot. Assume that the terrain is reasonably level and the first four grades on the surveyor's stakes read as follows: C-6.80, C-6.13, C-6.03 and C-5.90. The string line would be set for these cuts as follows. An 8 foot high string line should be used because a 7 foot string would only be .20-foot off the ground at the first grade. A 9 foot line would be more than 3 feet above the ground at the last grade. A line 8 feet above the flow line of the pipe is high enough to clear most ground obstructions yet low enough to step over easily. To set an 8 foot string line, add to each cut the number of hundredths and feet needed to total 8 feet. For example:

$$\left.\begin{array}{l} \text{C-6.80} + 1.20 = 8 \\ \text{C-6.13} + 1.87 = 8 \\ \text{C-6.03} + 1.97 = 8 \end{array}\right\} \begin{array}{l} \text{Even foot} \\ \text{eliminates hundredths} \end{array}$$

The same answers can be arrived at by subtracting the cuts required from the 8 foot string line. The method you use should be whichever is easiest for you. For the 6.80 cut, measure up the grade pin 1.20 feet.

Set a ruler on the surveyor's hub as shown in Figure 3-5 and measure up 1.20 feet on the ruler. Now set the straight edge against the 1.20 mark on the ruler. Hold the straight edge firmly against the ruler with the thumb and forefinger. Set a torpedo level on the straight edge and level the straight edge, holding it against the grade pin. Once the straight edge is level, have someone mark the grade pin. Be sure that if the top of the straight edge is held at 1.20 feet that the pin is marked from the top of the straight edge and not the bottom. Tie the string at the mark on the grade pin. Follow-

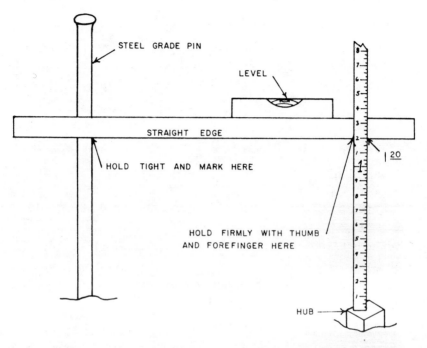

Marking grade pins
Figure 3-5

ing the same procedure at each station will produce a string line exactly 8 feet above the flow line.

If the terrain is steep, the grade line may have to be changed often to keep the string line from being too high or running into the ground. If there are large grade changes, the string must be tied in two spots on each grade pin and two measurements must be figured. See Figure 3-6.

Some inspectors do not think that an offset string line is accurate enough for checking final hand grading and pipe laying. Direct overhead line may be required. If this is the case, the grade line is set in exactly the same way, the only difference is that *T-bars* are used instead of grade pins. See

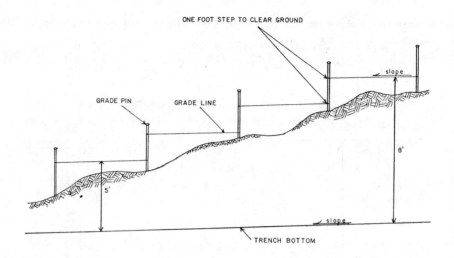

Stepped grade line
Figure 3-6

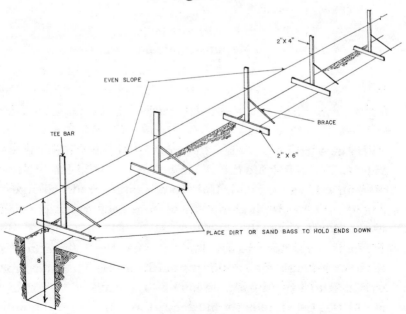

Overhead grade line
Figure 3-7

Figure 3-7. Instead of tying the string around the T-bar upright where the mark was made, a nail is driven in the board at the mark. Drive the nail just far enough to make sure it cannot be pulled out by hand. The ends of the T's must be weighted down with earth before stretching the string. Brace the T-bars with a 1x2 inch board so they will not be pulled over when stretching the string.

After all the bars are marked and nails driven at the marks, tie the string line. The string line can be tied directly to the nail on the first board with one wrap. On the next bar the string is pulled under the nail and wrapped around the bar once and back under the nail. Now pull the string tight from that bar to the next. When the string is stretched tight, flip it from beneath the nail to the top side. This will lap it over the bottom string and hold it from slipping. Note Figure 3-8. The same method is used on grade pins, except the string is looped twice around a grade pin.

Some foremen prefer a string line at the bottom of the trench so the man grading can check his own grade. To set a bottom line, the grade pins are driven at the bottom of the trench below the stations set by the surveyors. A straight edge is used as if you were setting the top line. Measure up from the grade hub 1 foot and on a level line across to the trench. Now measure down with a tape measure to the cut indicated on the surveyor's lath and mark the pin. See Figure 3-9. Do this at each station. When the string is set on those marks it will be one foot above the flow line of the pipe. Three men are needed to set a bottom line. One measures at the hub, one levels the straight edge at the trench, and one measures down with a tape measure and marks the pin in the trench.

All the measurements considered so far were to the flow

BOARD LOOP

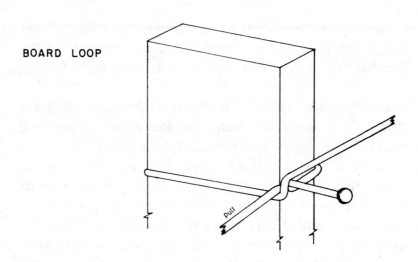

PIN LOOP

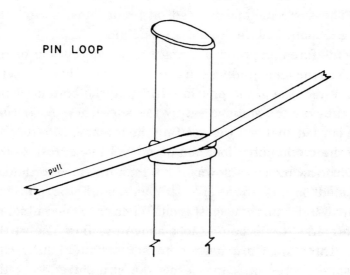

Board loop and pin loop
Figure 3-8

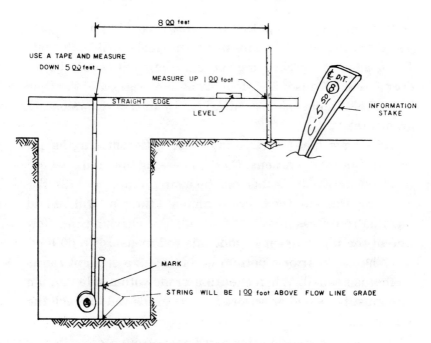

Setting bottom trench grade line
Figure 3-9

line of the pipe. The next thing that you must do is make up grade rods for final grading or pipe laying. Let's assume the specifications call for a .30 foot crushed rock bedding under the pipe being laid. The pipe is .20 thick from the outer wall to the inner wall. Once the bedding and pipe thickness are known, a grade rod for trenching can be made. This could be 1'' x 2'' or a 1½'' round wood or aluminum rod. Be sure the rod is long enough. Measure from the end of the rod up 8.50 feet and make a mark. The extra .50-foot is for the undercut, the thickness of the pipe plus the thickness of the bedding material: .20 + .30 = .50. The same procedure is followed when making up a grade rod for grading the bedding material except that the thickness of the pipe would be added to the 8 foot measurement, making it 8.20 feet. The grade rod for checking the pipe flow line would be an even 8 feet because the 8-foot string line was set to the flow line. The flow line is the grade given by the surveyors on the grade stakes.

You have made three rods three different lengths for three different operations: (1) For checking the grade behind the trencher or hoe, this rod measures 8.50 feet. (2) For checking grade for the bedding material the pipe will be laid on, this rod measures 8.20 feet. (3) For checking the flow line of the pipe after it is laid, this rod measures 8.00 feet.

When a T-bar or a bottom line is used, a straight rod is all that is needed. When checking grade with a side line, an arm must be nailed or screwed to the grade rod to reach the side string.

Setting a string line for any above ground operation is done the same way with one exception. When setting a string line for a paving machine, self-grading curb machine, or checking any above ground trimming with a hoe, there will be fills on some of the stakes. If the surveyor has a fill indicated rather than a cut, use the following method for setting the string. Assume string is being set 3 feet above the desired grade. The correct amount is added to each cut to get an even 3 foot reading. Then suppose a fill grade is encountered. The surveyors' stake shows a 1 foot fill to grade. If a 3 foot string line is being set, just add the fill grade given, regardless of what the fill is, to 3 feet. Remember that even though the grade is being set for a surface grade, you should keep four grade pins set up at all times so any mistakes are apparent.

When setting a string line for curb boards or anything

The concrete machine is about to start a radius to the left.
Notice the grade line stops at the radius and a ½ inch plastic
pipe is used so the machine will make a smooth turn around
the radius. The plastic pipe dragged at the right acts as a
tape measure. It is 20 feet long and marks the 20 foot
intervals where expansion joints must be set.

Figure 3-10

where the string is to be placed at a finished grade, the
measurements given by the surveyors are used as is. If the
surveyor's stake has a 1 foot cut to the top of the curb,

measure out the offset given and down from the level straight edge one foot and mark the pin. If the ground in front of the hub to the curb is higher than the hub, use an extension on the straight edge. The amount of extension added must be computed with the cuts and fills given on the surveyors' stakes.

When setting a string line for forms of any kind, set the pins and string up so the pins will be on the outside of the boards. Corners are difficult to make when setting up a string line for a self-grading curb machine. Many pins must be set to keep the string flowing evenly around the turn. Otherwise the line sensor will cause the machine to make a very irregular curve. This problem can be solved by using a ½ inch plastic pipe in place of string on the corners. This will make the corner much easier to set, will use less pins, and help the machine turn out a smooth corner. See Figure 3-10.

On a sewer line project, when the locations of the sewer services are staked, the surveyor usually will not give a cut grade. When the sewer service is dug, the trencher or hoe operator may want a grade line to follow. If this is the case, it can be done without assistance from the surveyors. Offset a pin from the service Y on the sewer main and do the same at the service location stake set at the property line by the surveyors. From the bottom of the pipe measure up and over 6 feet to the grade pin. See Figure 3-11. Assume the trench is 5 feet deep and we have measured up six feet and over to the pin. Compute the amount of minimum fall required in the specifications and raise the string at the property line that amount. This is done by shooting level from the front pin at the Y to the back pin at the property line. Then raise the back string above the level the amount of fall desired and

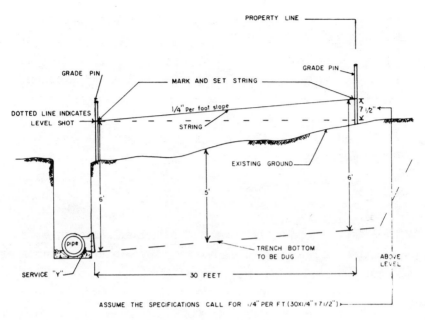

PROPERTY LINE

GRADE PIN

MARK AND SET STRING

GRADE PIN

DOTTED LINE INDICATES
LEVEL SHOT

1/4" Per foot slope

STRING

7 1/2"

EXISTING GROUND

6'

6'

5'

pipe

TRENCH BOTTOM
TO BE DUG

SERVICE "Y"

30 FEET

ABOVE
LEVEL

ASSUME THE SPECIFICATIONS CALL FOR 1/4" PER FT (30x1/4" = 7 1/2")

**Sewer service grade line
Figure 3-11**

mark. If the ground rises sharply from the Y to the property line, tie the string above the ground the same distance at both the back pin and the front pin. The service line will then have approximately the same fall as the ground slope.

When setting any grade line, be sure to check every figure. Avoid any distraction while figuring or marking grades. It is good practice to use a pencil and paper to figure grades. A common mistake is to figure grades from a string of cuts and then come to a fill and figure it as a cut. Look at each grade stake separately and read it for what it is.

4

Excavating Subdivisions

This chapter will cover the basic steps to follow in excavating and grading subdivision streets and lots. The job should proceed in the sequence outlined in this chapter. As in most excavation work, planning is important. Never let the equipment stand idle while grades are being set or while a decision is made on where to start the next cut or fill. Try to solve problems before they hamper production. Study the plans and stakes carefully before starting. Take the time to analyze the type of equipment required. This will eliminate reduced production that results from poor equipment balance. If using a paddle wheel scraper, know the soil condition. Determine whether a paddle wheel with ripper teeth can cut the material or a tractor ripping must be added to help. Check the soil moisture and distance to the water source. This determines the number of water trucks needed and the truck size. Remember, the size of the equipment should be determined by the size of the job.

Problems may arise that are not covered in this chapter. Most of these are covered elsewhere in this book. Check the index for additional information.

The surveyors will lay out the subdivision property by running a row of stakes down the center of each street. On each stake they will note that the stake marks the center line, write the station number, and give the cut or fill to the finished road grade. At the base of each information stake there will be a hub driven in the ground. From this hub the grades shown on the information stake will be taken. The hub will usually have a small tack somewhere on the top surface which is the reference point on the hub. From that point the surveyor computes the grades at that station.

The lots are staked at each corner of the building pad and usually at the back property line. If the lots are large, they may have additional intermediate stakes. Each ·stake will have written on it the information needed to cut the lots to grade at that point. In some cases one stake and one hub can be used for two adjacent lots. In this case one side of the stake will read *lot 22 South, 10 feet,* and the amount to cut or fill. On the back side of the same stake will be *lot 23 North, 10 feet,* and the cut or fill. It is very common to have a fill for one lot and a cut for the adjoining lot. In most cases the surveyors tie different color ribbons to the tops of the stakes so the construction crews can easily distinguish the lot stakes from the street stakes. Front and back lot pad stakes may have offset distances on them. Figure 4-1A shows front lot and street center line staking.

To start excavation, the area should be stripped of all vegetation by either disking it under or scraping it off with earth moving equipment and then dumping it in a desig- nated area. Note Figure 4-2. The strippings are usually dumped at the back of the lots. It is very important that no

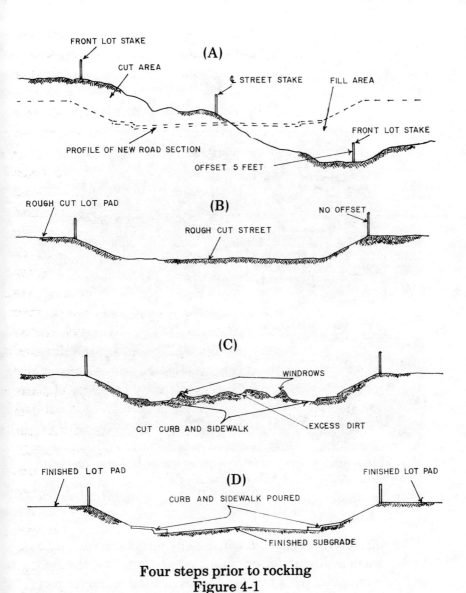

**Four steps prior to rocking
Figure 4-1**

vegetation be placed in the building pad area because most
anything that grows is unsuitable for compaction.

The self-loading scraper is stripping grass before any cuts
or fills can be made
Figure 4-2

A soils engineering firm will usually direct the contractor
to the areas available for dumping vegetation. The engineer
will also check lot pads for clay pockets. If the engineer feels
a lot pad has too large a clay pocket, he may direct the
contractor to remove the clay and compact suitable soil back
into that area. The soils engineer will usually test all fills at
various levels as they are being filled.

After the entire job site has been stripped of all vegetation, the grade setter can set his street stakes. The street width is indicated in the road section on the plans. The grade setter should run two rows of stakes on each side of the street at the back of the proposed curb or sidewalk, parallel with the center line stakes and at the same stations. For example, if the plans indicate the road is 48 feet across from the back of sidewalk to the opposite back of sidewalk, the grade setter would measure 24 feet from the center line and drive a stake at a right angle to the center line. This is done at each station and on each side of the center line. This defines the street area to be cut before the slope starts up or down to the lot pads.

The grade setter may also have to transfer the center line grades set by the surveyors to his stakes at the back of the proposed curb or walk. If the excavation crew is experienced, the lath will not have to show the elevation. The grade at center line is the only grade indication needed. The grade setter can check grade at each station and tell the scraper operators how much cut is left to make. The grades set by the surveyors indicate the height to the finished road surface. The grade setter therefore must add the thickness of the road section to all cut grades and subtract the same thickness from all fill grades. By doing this, all the cuts and fills become subgrade cuts and fills rather than finished grades.

Remember to consider underground utilities while setting up subgrade cuts and fills for a subdivision street. Placing drainage and sewer lines and hauling material in for the initial backfill over the pipes will create a surplus of soil. If this excess is not compensated for during the street excavation, it will hinder the crew that returns to cut the curb grade after the utilities are in. The foreman must determine the amount of excess anticipated from the utilities and undercut the subgrade to allow for that excess. A good rule of thumb to follow is to undercut the center line the amount required for the crown. For example, if the road is to be 24 feet wide from lip of curb to lip of curb with a 2 percent crown at the center line, add .24-foot to the subgrade cuts at the center line. Cut a flat grade and allow for the excess utility dirt to make the crown later. See Figure 4-1 B, C and D. If the soil is very sandy, this practice of ignoring the crown will not apply because sandy soil shrinks more when compacted.

If the underground utilities are more than 24 inches in diameter, more undercutting may be needed, depending on the amount of structure backfill specified. If the structure section of the road is much thicker than the curb, the flat subgrade cut must jump up at the lip of the curb so as not to undercut the curb subgrade.

The actual excavation can start once the grade setter has all the street stakes and grades in place with the proper figures added to cut the subgrade. Blue survey tape should be tied to all fill stakes to indicate the point the fill should reach.

Excavation should start in the following manner. The cut and fill stakes should be studied closely and a cut and fill pattern figured out to eliminate any long hauls. Start at the front lot stakes and work toward the center of the street. Keep the cuts as even and level as possible. With each pass the scraper should move more toward the center of the street. The steepness of the front lot slope will determine the amount the scraper will move toward the center line on each succeeding pass. The excess dirt should be hauled to the nearest lot or street area needing fill.

If the front lot stake shows a fill and the street stake requires a cut, make the street cuts and continue filling

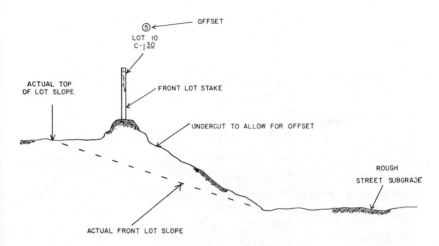

Front lot sloping
Figure 4-3

along the front of the lots until the fill grades marked are reached. If the front lot stake has been offset toward the street, the lot slope on the street side of the stake should be undercut. The amount to be undercut depends on the offset distance and the steepness of the front slope. See Figure 4-3. Once the lot fills and cuts have been made, the front lot stakes should be set at the actual front lot pad corner. Then a scraper can straddle the ridge of dirt left and blend the front lot slope with the lot as shown in Figure 4-4.

Fill the front portion of the lots first. Then work across the lot from the front stakes to the back property stakes. Attempt to set up the earth moving operation so the spread can work from one cut area to the next. The fill areas between cuts should be completed as the dirt spread moves ahead. This eliminates any doubling back once the area has been left.

While rough grading the lots, it is desirable to leave them .10-foot above grade. There will be some shrinkage

A scraper cutting a front lot slope
Figure 4-4

during the lot trimming and rolling and trimming is much faster if the grader operator has a slight excess to work with.

If, at any time during excavating the stakes seem to be off grade or it looks as though dirt from the cuts is not going to balance with the fill needed, call the engineering firm that staked the project. They will be happy to recheck any stakes that seem to be off. Any experienced survey crew has made a mistake on a station or two. Also it is not uncommon for surveyors to adjust a few lot grades to get the dirt to balance. The important thing is to watch for either of these

A grade setter is shooting a lot pad grade ahead
of the grader
Figure 4-5

two problems so they can be corrected early. The front lot
slopes can be left slightly concave to accept the excess lot
trim dirt.

After all the rough cutting has been completed, the lots
should be fine graded to the tolerance called for in the
specifications. If the specifications call for a lot tolerance of
.20-foot plus or minus, the lot grading can be expedited by
averaging the cuts and fills. For example, suppose every

second lot rises .10-foot. The eighth lot would then be
.40-foot higher than the first. Raising the grade on the first
lot .20-foot and lowering the grade on the eighth lot .20-foot
will create a level grade through the eight lots but still stay
within the .20-foot tolerance specified. This speeds up lot
grading because the grader operator has a longer run
without a grade change. Figure 4-5 illustrates this situation.
If care was taken during the excavation and rough grading of
the lots, the fine grading can be completed by using a grader
and rubber tire roller. The small amount of excess dirt
generated can be "lost" at the front of the lot between the
building pad and street.

When the lot grading has been completed, underground
utilities are placed. Then the surveyors return to stake the
curbs and walks. If there are to be no curbs, ditch and street
grades should be set. All the grades given will be finished
grades. If there is a ditch rather than curb, the ditch and
road grade should be cut. If curbs and sidewalks are
required, a grader and grade setter will cut the curb and
sidewalk grades. The grade setter must add to the cuts and
subtract from fills the thickness of the concrete curb or walk.
Remember to take into account any rock material that might
be called for under the curb. Refer again to Figure 4-1, C.
Once curb and walk grades have been cut and the curbs and
walks poured, the street compaction and fine grading can
proceed.

The grade setter will take his final street grades from the
curbs by computing the percentage of slope to the center
line. For example, if the street is 30 feet wide and has a 2
percent crown, the center would be three tenths higher than
each curb lip at the finished grade.

The grade setter can set his grade in three ways. He can
shoot it with his eye level, stretch a string from lip of curb to
lip of curb and measure down, or use three laths and sight
across the tops. The three laths method is called *swedes* and

the center lath is shortened or lengthened to give the desired crown. All three methods are accurate. The string line and swedes are faster but they take two or three men rather than one. In some cases the specifications call for the surveyors to restake the grade each time.

By this time a small excess of dirt will have built up from the curb grade cutting and excavation for utilities. This dirt can be dumped on any low areas at the front or back of any lot. House pads have already been fine graded and these grades cannot be changed. When the excess dirt has been removed, the road should be approximately .10-foot above grade to allow for shrinkage during compaction.

For good compaction, the road should be ripped up six inches deep, watered, mixed thoroughly and then compacted. Usually a compaction test of 95 percent will be required for the top 6 inches. After the compaction process has been completed and the road reshaped by the grader, center line hubs can be re-set. Take the grades from curbs by using swedes, reshooting, or string lining. Now the street can be fine graded and rolled.

The subgrade is now ready to be paved or rocked, whichever is required in the specifications. See Figure 4-1, D. If it is to be paved, no further grade setting is required. If, however, rock is to be placed, the grade setter should set a row of short laths about one foot high down the center line. Mark each stake for the finish rock grade. This will give the grader operator something to work to while he is spreading the rock.

After the rock is in and compacted, the grade setter will set hubs or the surveyors will reshoot the grade at the center line for the fine grading. When the grading has been completed and compaction tests passed, the rock is oiled and the road paved without additional grade setting.

Two important items should not be overlooked: first,

when a deep cut or fill is encountered, the surveyor's stakes must be raised or lowered with the fill or cut by the grade setter and the grades adjusted accordingly. If a cut is no more than about four feet, the equipment can work around the stake, leaving it on a mound until the rough grading has been completed. Then the stake can be reset and the mound wiped out. If it is the last pass in the street, the mounds can be cut and the stakes not reset. Second, when the curb is poured, the location of water services and sewer services must be marked on the curb so they can be located later. If there are no curbs, the locations should be marked with a 2 x 2 redwood stake.

Clean-up follows once the subdivision roads have been paved. A grader should dress up dirt behind the curbs or walks and also any rough areas that have been left at the front and back of lots. All stakes and debris left during construction must be hauled off. All the manholes, water valves and sewer clean-outs in the streets and easements should be raised. These should have been measured out and marked on the curb before being covered during the subgrade operation. After the manholes, water valves, and sewer clean-outs have been poured and paved around, the subdivision should be complete and ready for the final inspection.

5

Highway Grading

This chapter explains how a road job is staked and the steps used to complete the job. Problems may occur that will cause the order of work to be altered slightly from what is described here. This chapter describes mainly excavating procedures. Compaction, subgrade preparation and rock grading are covered in separate chapters.

Highway construction is one of the most difficult excavation jobs because of the large number of changes in grade that occur between the right-of-way line and the center line of the road, island, or median. In most cases there is no concrete curb to aid the final grading.

The first cuts will probably be for a ditch at the top of the slope, followed by a slope cut to another ditch. From the bottom ditch there may be a fill to the shoulder grade and then a slope to the center line of the road. If there is an island or median between lanes, those grades will also be indicated on the information stakes. Study the typical road

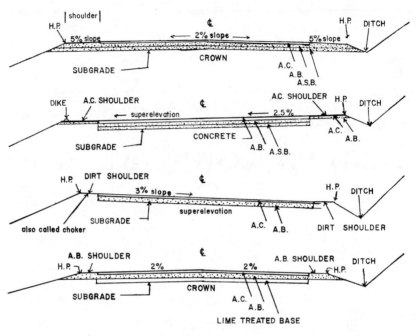

Road sections
Figure 5-1 (A)

cross section in Figure 5-1.

These grade changes are less imposing if the men doing the work can visualize what the finished grade should look like. This will help keep them from getting confused by all the stakes the grade setter must set to lay out the work. The foreman and grade setter should watch the excavation equipment carefully on highway work until the men become familiar with the fill and cut areas.

When staking out a highway job, the surveyors usually run a row of information stakes and hubs at the right-of-way line on each side of the road. The right-of-way line on each side of the road is usually the limit of the construction area. No work should be done and no equipment should travel

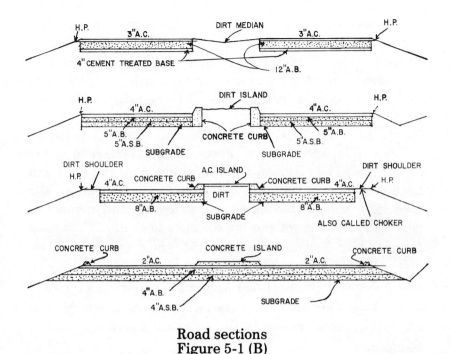

Road sections
Figure 5-1 (B)

beyond the right-of-way stakes without a property owner's permission in writing. If the grade stakes are not on the right-of-way line, the right-of-way line will usually be staked separately at 100 foot intervals.

On the stakes at each side of the road are written grades for every change in elevation from the stake to the center line of the road. These stakes will be at 50 foot intervals or less and parallel the road. The cuts and fills start from the RS point and grade given. If the road is wide and has a median down the center, there will probably be a row of stakes there also. If during clearing the ground level at the RS point was lowered, start the fill closer to the information stake to compensate for the cut that was made. *Example:* The fill was to be a 2:1 slope starting from the RS point given on the stake. In clearing the grass one foot of dirt was also removed, leaving the RS point 1 foot low. Compensate for this by setting a stake 2 feet back of the RS point given on the information stake. At a 2:1 slope this will yield the grade and distance given by the surveyors when the fill is 1 foot high at the RS point.

On the front of the survey stake the surveyor will list all the cuts, fills, and distances needed to build the road to the center line. The stake on the opposite side of the road will have the information needed for the remaining half of the road. The back of the stake will give the station number where the stake is set and the distance to center of the road from the hub. The distance from RS to the center line will be on the stake front. On the sides of the stakes appear the fill or cut slope, such as 2:1 or 1:1, and the percentage the road slopes from the centerline to the *HP* or the shoulder. Also on the side of the stake will be the elevation above sea level at the hub.

If the road grade from the center line to the shoulder is rising at 2 percent, the stakes will read +2%. If the road slopes down from the center line to the shoulder of the road, the stake will read -2%. See Figures 1-1 and 1-3. In some cases the survey stakes may not have all this information. It depends on the surveyors and local practice. When the grade stakes do not have all the percentages or information needed examine the plans for the remaining information. If you can read the grade stakes, you should have little trouble completing the necessary grading as described in the remainder of this chapter.

First clear all the vegetation and trees from the road area and dispose of it as the specifications provide. The grade setter uses the grades and distances supplied by the surveyors on the information stakes to set out his stakes. The operators then know where the cut and fill areas are without stopping to read the surveyor's stakes. The grade setter will place cut stakes first. This means that he will put a stake where the slope will start downward to the road grade. On this stake he will write the cut to be made, the rate of the slope, and the distance to the bottom of the cut slope. Some grade setters indicate only the vertical feet of cut and the horizontal distance to the cut bottom. If a grade setter has experienced operators on the equipment, he can give them the rate of slope. They will know how much to move out with each pass they make.

The grade setter should then move to the fill area and set a stake at the bottom outside edge of the fill to be built up. This is the *toe of slope* and is one of the RS points given. On the stake at the toe he will put the same information as on the cut slope. He will write the feet of fill, the distance to the top of the fill, and the rate of slope. The stakes set by the grade setter are crows feet with no hubs needed. Refer back to Figure 3-1 (A). If there are any ditches to be cut at the top of the cut slopes or bottom of the fill slopes, these must be staked and cut before cutting or filling the main slope.

Once all the outer ditches are cut and the job stripped of all vegetation, the main earthwork can begin. The equipment operators must know the rate of slope of any fill or cut being made. It is very important not to undercut a cut slope or underfill a fill slope. Each time the fill slope rises about four feet, the grade setter should set another row of stakes along the full length of the top edge of the fill. These stakes should have new fills and distances written on them to indicate what is needed from that point to the top of the fill. To set a new row of stakes on a fill using a 2:1 slope, the grade setter will measure out 8 feet and drive a lath. He then will measure up 4 feet from the previous lath's grade mark and shoot level with an eye level to the lath 8 feet away.

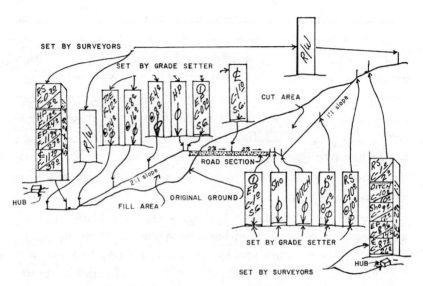

Staking for typical cut and fill section
Figure 5-2

Then at that level point, he will draw a horizontal line. That is the mark the fill must reach.

Setting these fill stakes in the slope periodically helps both the grade setter and equipment operators keep track of the progress of the fill. Figure 5-2 illustrates the markings these progressive fill stakes might carry. If he finds the fill slope is getting too steep, the grade setter can tell the operators to move into the slope more. On the other hand, if the fill is coming up too flat, the operators note the stakes and fill more to the line marked by the grade setter.

It is a good practice to overfill the fill slope slightly so that it will be about right when compacted. Usually a .50-foot tolerance is allowed on a fill slope. Keep this tolerance on the plus side to avoid problems later when it will be more difficult to build the slope up to what is required.

The grader is making the first sloping pass in the process of cutting a 5 foot bank to a one to one slope. Notice that a smooth grade was established at the bottom of the bank to make excavation of the bank easier. The wheels are slanted away from the bank to put more pressure at the cutting edge
Figure 5-3

The surveyors should always be notified if the information on one of the stakes does not seem to match the others. The surveyors make mistakes occasionally. The sooner they are notified of an error, the sooner your work can be finished.

Two sloping passes have been made. The grader is now
making a final trim to cut the toe of the slope back to where it
should be. At the same time the bottom grade will be
smoothed so the final slope pass can be made
from a level base
Figure 5-4

In the cut area keep the slopes shallow enough so they
can be reached by the grader. The cut slopes must be trimmed to grade as the cut proceeds downward. If a grader is
used, the cut can be taken down about five feet before it
must be trimmed by the grader. See Figures 5-3, 4 and 5.

In this photograph the slope has been trimmed for the final time. The grade setter must measure the top of slope distance and the bottom, or toe of slope distance. If the cuts coincide with the distance on the stakes the slope is finished

Figure 5-5

The grade setter should check the grade ahead of the grader operator and let him know how much is to be trimmed. To do this on a 1:1 slope, measure from the top of the slope or previous stake out five feet horizontally. Then, holding the ruler horizontally, drop a pebble from the five foot mark. At the point the pebble hits the slope below, set the ruler vertical and shoot level back to the top of slope or previous stake. If the ruler reading at that point is less than five feet, more of a cut is needed. If it is more than five feet, too much has been trimmed off.

On most jobs there is a tolerance of .50-foot on the slopes. When the slope has been trimmed to grade, the grade setter should set stakes on the slope with a line indicating *at grade*. He should indicate the remaining cut distance left to go. This procedure is continued over and over until the cut reaches the toe of the slope called for on the grade stake set by the surveyors. The grade setter does not need to take his grades from the original survey stake each time. He can take the grade from his slope stakes each time and change the elevation and distance accordingly. See Figure 5-2.

Sometimes in the cut area there is a ditch on each side of the road that must be cut lower than the finished shoulder of the road. In this case the initial cut should be to the finished shoulder grade. After the dirt has been removed to this point, all the lower grade elevations such as ditches should be cut.

If there are shoulders which are to remain higher than the subgrade (usually called *chokers*), this slope or vertical cut should be trimmed. Remember that it is the finish grade the surveyors are staking and the road section thickness must be subtracted when staking subgrade for cuts and fills. Leave dirt shoulders or chokers .10 or .20-foot high so they will be easier to finish after the road has been paved. Many of the roads you work on will have dirt shoulders and the shoulder finish grade will be higher than the road subgrade. Once the highest grade has been reached in a cut area, move the equipment to the inside of the shoulders and continue cutting until the road subgrade is reached.

In fill areas the fill is brought up to the road subgrade and then shoulders are built up and trimmed to the width specified. Leave the road subgrade low enough so that when the shoulders are trimmed the excess dirt brings the subgrade up to the right grade. This eliminates extra trimming work on the subgrade. If the road has a dirt filled island or median, handle it in the same manner as a shoulder or choker.

The grade setter must be alert when the cut operation is approaching the shoulder or island grade and the fill operation is nearing the subgrade. The operators need plenty of stakes and guidance so they don't cut or fill too much or in the wrong areas. Make sure each operator has a good picture in his mind of what he is building. This can save the grade setter, foreman and contractor a lot of unnecessary trouble.

In some cases the fill slopes must be rolled and trimmed once they are up to grade. You can compact the fill slope while the fill is coming up if a cat and sheepsfoot roller are available. Have the operator run the sheepsfoot down the fill each time about 8 feet of fill have been deposited. This way, the sheepsfoot can be used on the slope while the dozer remains on the top of the fill. If a self-propelled compactor is used while the fill is being built up, a sheepsfoot roller must be used later. Use a sheepsfoot roller attached to a cable and crane if the fill is too high for a cat pulled sheepsfoot to reach down or push up. A dozer can then trim the slope to the required grade.

The road is ready for subgrade work once the slopes have been compacted and trimmed and the shoulders are built and cut off vertically on the road side. The surveyors may then set bank plugs for the final grading operation. See

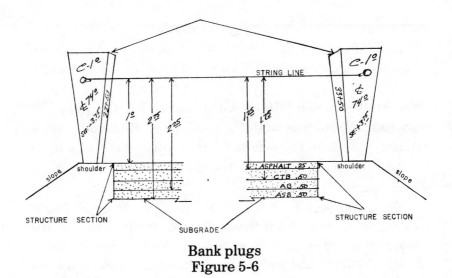

Bank plugs
Figure 5-6

Figure 5-6. If bank plugs are set, no other grades are needed for subgrade, rock, or pavement grade. If the surveyors do not set bank plugs, they will set grade hubs in the subgrade after it has been rough graded and once again when the aggregate base is ready for fine grading.

The subgrade of the road should be trimmed to approximately .10-foot high before being ripped up and compacted. After compaction, the subgrade is trimmed again, this time to the tolerance in the job specifications, usually to .08-foot plus or minus. Be sure the entire width needed for base or pavement has been excavated and compacted. Never let chokers encroach into the base area.

If the subgrade passes the compaction tests, the road is ready for placing road rock and paving. The thickness of the rock, pavement to be placed, and pavement type vary greatly. Later chapters cover this work. When paving is complete a final fine grading is usually required.

The shoulders, islands, and ditches should be trimmed of any excess dirt. Shoulders must be rolled and the ditches

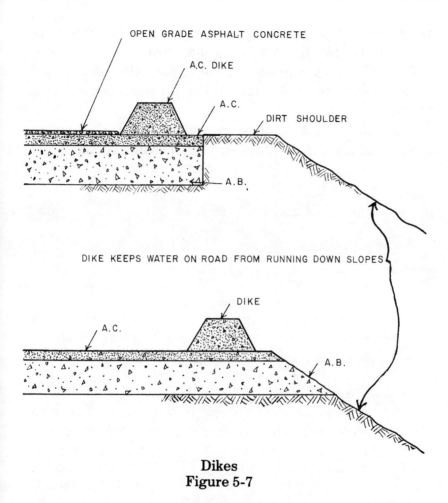

Dikes
Figure 5-7

shaped. The road can then be washed. If dikes are to be placed, now is the time to place them. If an open grade asphalt is called for on the top lift (road surface), be sure the dikes are in place before the open grade asphalt is applied. Open grade asphaltic concrete (sometimes referred to as popcorn) is coarse asphalt concrete, usually having ¾'' aggregate with very little fine material added. If a dike such as in Figure 5-7 is placed on it, water will seep under the dike and out of the back side. The dike is intended to prevent water from running off the road and down the slope. A dike placed before open grade asphalt is put down on the road will collect water as it runs off the road and carry it to a drain.

The final step is to place road signs, paddles, and a fog seal of oil if required. Road striping finishes the job.

6

Planning Excavation

This chapter will discuss the most practical excavation methods and the preferred procedures for equipment operators. In general, you will operate most efficiently if you begin by imagining that dirt must be moved with a stick. You must scrape the dirt from each high area to the closest low area without carrying it. Follow this simple rule whenever possible to establish the most efficient method possible for moving dirt.

The prime objective in planning excavation work is to make dirt move the shortest distance possible from cut to fill. Your plan is efficient only if all the dirt on a project can be moved from cut to fill without traveling over a previously cut or filled area. This is not always possible, but it should be your goal.

Assume you have a road job as shown in profile in Figure 6-1. Your job is to find the most efficient excavation plan.

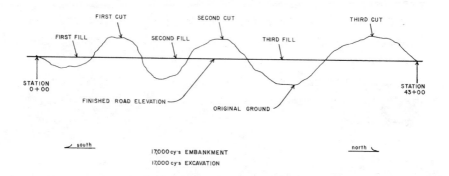

Cut and fill profile
Figure 6-1

The job quantities are 17,000 cubic yards of fill (embankment) and 17,000 cubic yards of cut (excavation). This is a balanced dirt job because all the excavation will be used as embankment with no excess to remove. The dirt spread should begin on the south end of the job, the left side of Figure 6-1, because that is where the smallest fill areas are located. When the two smaller fills are made, the first cut area will also be cut to rough subgrade.

Subgrade is the portion of road bed on which pavement base and sub-base is placed. Rough subgrade is grade level cut using crows feet. Finished subgrade is trimmed level with grade hubs driven in the subgrade. Refer to the road cross sections in Figure 5-1 for a subgrade diagram.

Next, the grader can start cutting ditches and trimming the rough subgrade while the earth moving equipment finishes moving the remainder of the second cut material to the third fill area. When the second cut area has reached subgrade, the scrapers will move to the third cut area and finish filling the third fill area to grade. This will finish the excavating. Only one scraper has to remain on the job to move any excess dirt generated by the grader.

The amount of excess dirt generated by the grader while cutting ditches through cut areas should be estimated. The rough subgrade should have been undercut during excavation so that the ditch dirt can be used as fill on the subgrade. This will leave a small amount of trim for the scraper to double back to pick up. If this is done correctly, a smaller scraper may be moved in sooner for rough subgrade balancing and the job cost will be lower.

If chokers are being built in the fill areas, the rough subgrade should be left low enough to use the excess dirt when the chokers are trimmed. A choker is a dirt shoulder built up higher than subgrade level. The scraper operators should always leave their cuts as level as possible.

If a push cat is being used, set the scraper in position for the next cut close to the dozer, this way the dozer doesn't have to travel too far. A good dozer operator can direct the scrapers with hand signals to the area he wants cut next. If the starting place is the first cut area in Figure 6-1, the scrapers should be headed south while cutting and the cut should start at the top of the hill. The operator should cut as deep as possible at first. As he starts down the hill he should keep raising the bowl slightly. Doing this on each cut will eventually flatten the top of the hill, giving the operator a good flat surface from which to work. Then a smooth light cut can be made on a level plain, making the loading easier. Never load going up hill if at all possible; it is very difficult and slow.

Material placed in the first fill area in Figure 6-1 should be started at the lowest point. From the lowest point the fill should be brought up in smooth, level lifts until the desired grade has been reached. The scraper operators are responsible for keeping the fills and cuts level. The push cat, ripper cat, or compactor should be concerned only with what they do best: pushing, ripping, and compacting. The dozer on the compactor should be used to mix the fill dumped and to keep the moisture distributed evenly.

Scraper operators have been known to gouge up the cut area and dump piles in the fill areas, relying on the dozer or compactor to level the piles. The operator that thinks he is saving time by this quick gouge and pile method is badly mistaken. Taking a smooth cut while loading and spreading the dirt evenly while dumping saves time in the long run. Keeping the cuts and fills smooth and level will be much easier on the operators and add to the production efficiency.

Avoid spinning the tires when loading a scraper. Spinning the tires causes excessive tire wear. After a few passes the operator should be able to tell by the sound of the engine and the ground speed just when the tires will start to lose traction. When he feels this is about to happen, he should raise the bowl slightly until the engine strain eases and the speed picks up again. A very skilled paddle wheel scraper operator can take a light cut in second gear and load faster than in first gear. If loading in first gear is easy, try second gear and time the loading period. If the loading time is faster, try third gear with a lighter cut yet. If that is faster, stay with it. If it is slower, drop back to second gear.

There is no rigid standard operating procedure for any one piece of equipment because soil conditions vary so greatly. But regardless of the equipment used, try to develop a faster, easier way of getting the job done. The foreman should gather his operators before starting any excavation and discuss the travel pattern he wants used from excavation to embankment. Tell the operators where the excavation and embankment should start. Point out soft areas that should be brought to the operator's attention. Explain the best method of working once the work begins. Encourage the operators to suggest a better plan of operation if they see one.

The most important thing a foreman should provide his scraper operators is a good, smooth haul road. It is probably the single most important factor in achieving good production. Even the best scraper operator's talent will be wasted if he must haul over a rough road. On long hauls a grader and water truck should make a pass over the haul road periodically to keep it in good condition. This will minimize the scraper cycle time and reduce the man and equipment hours needed to finish the job.

7
Drainage Channels

This chapter will explain the preferred method of staking and excavating drainage channels. It will point out the problems that might arise and the best ways to minimize these problems. Many problems can be avoided in the beginning if you select the right equipment for the job. Equipment too big for the work or too much equipment will result in lower productivity every time. Plan your equipment needs carefully on a channel excavation job.

A channel is usually staked with two rows of stakes, one on each side of the channel. The stakes should be offset far enough from the top edge of the channel slope so that they do not interfere with the equipment. Marked on the stakes will be the fills or cuts that were computed from the hub at each stake for the following distances: the distance out from the stake where the cut begins (top of slope) and the bottom of the cut (toe of slope), the distance to the center of the channel. The stakes are usually set every 50 feet.

When cutting a channel, the foreman should determine

the best locations for the ramps in and out of the channel. In most small channels there is not enough space for one piece of equipment to pass another. A good route should be worked out so the equipment returns to the entrance ramp as directly as possible. Narrow channel work can be very slow because grading and excavating equipment cannot work simultaneously. Still, operators must be very careful not to undercut the slopes. When excavation along the slope has reached a depth of about five feet, the slope should be trimmed with the grader. On a narrow channel the scrapers might have to wait for the grader to finish the sloping pass before they can get by. If this is the case, it is essential that the grader operator be both accurate and fast so as not to hold up the scrapers longer than is necessary. Channel cutting can be very difficult for an inexperienced contractor.

In some cases the dirt dug from one section is needed in a preceding section of the channel to build the sides before cutting. Once the channel has been dug the ramps can be removed. The sides where the ramps were should then be trimmed with a hoe or Gradall.

Widening an existing channel can be a real problem. It can be approached in any of several different ways. The first requirement is to re-route the water around the construction area. If the bottom is muddy and grassy, a hoe or dragline must be used to clean it out down to stable material. Then dirt can be hauled in from a borrow area and the channel filled to the top. Earth moving is then done as if the channel were being given a new alignment.

If no borrow dirt is available, the work will be very slow because excess soil from slope trimming must be used to fill the existing slope voids. The only advantage in this case is that no equipment will be traveling on the channel bottom.

Therefore the bottom can be left soft. The slopes must be trimmed with a hoe or Gradall. In areas where slope filling is needed, a bench or terrace must be built and then the slope overfilled so that the hoe will have only trimming and no filling to do. This is an expensive way to do this work. The hoe must first trim the dirt. Then it must be hauled to the slope being filled. Finally it is re-cut by the hoe. The dirt is actually being handled three times.

As you can see, channel cutting can be difficult. In most cases the absence of working room and the soft bottom cause most of the problems.

When widening an old channel where the cut on the top is not wide enough for the scrapers to get started, a cat should be used. The cat can start by dozing the slope out to a point where it is wide enough for the scrapers to begin work. This same method must be used in widening cut slopes on road jobs or starting narrow fills. The dozer operator, in bringing down the cut slope, must be careful not to undercut the slope. If the channel is going to be lined with concrete, care should be taken while cutting the slopes to keep them as straight as possible.

On a large drainage channel where water is a problem, it may be necessary to excavate in two stages. Use a scraper for the first stage until the bottom becomes too wet and soft. The first stage slopes will be trimmed with a grader. Then a hoe or dragline will have to finish the job. When ground water is a problem, a well point pumping system may be needed to control the water level. This will help you in two ways. First, it will allow the scrapers to work longer before the hoe or dragline is needed. Second, when the hoe or dragline finally is required, the well points will reduce the amount of water in the channel, making excavation and grade checking easier.

Planking or sheeting may be needed under the tracks of a hoe or dragline to prevent the equipment from getting stuck. If the bottom is firm enough, use a small dozer with a slope bar to trim the slopes and the bottom to grade. This will relieve the hoe or dragline of this task and improve production.

There is usually little clean-up left to do when the channel excavation has been completed. Re-setting existing fences or constructing new fences will be the only item required to finish the job.

8

Unsuitable Material

Unsuitable material is any soil or aggregate that has absorbed water to the point where it does not provide adequate support for a road section. It is usually recognized by a rolling movement of the ground as the equipment runs over it, somewhat like the way a waterbed ripples. A surface that moves this way is said to be *pumping*. Any time you remove unsuitable material, keep an accurate record of the quantity of material removed and replaced. Any unsuitable material removed below subgrade level will usually be allowed as an extra charge. But be sure to read the specifications before removing any unsuitable material and discuss it with the inspector, engineer, or owner. If the specifications or bid items do not cover unsuitable material, do not remove it until an agreement is reached on payment. Some project specifications require the contractor to remove unsuitable material at his own expense. On a project where you must remove unsuitable material at your own cost, it's going to save you money if you can distinguish unsuitable material from suitable material. Distinguishing the two soil conditions with accuracy takes a great deal of experience and should not be attempted by an inexperienced foreman or superintendent. Seek advice when you are in doubt.

Earth with a heavy clay content may move under the weight of rollers as though it were unsuitable material. It may actually be quite stable for most purposes. Test for unsuitable soil as follows: press down with all of your weight on the balls of your feet. If a small area around the soles of your shoes sags from the weight, the soil should be considered unsuitable. If the ground smashes under your feet but the area around your footprint does not sag, it is usually firm enough to hold aggregate or pavement.

When removing unsuitable material, make sure you remove enough to expose good firm soil. If a firm bottom cannot be reached, let the inspector or owner make a decision on what should be done next. If four feet of unsuitable material has been removed without reaching a stable base, you can recommend the following method. It will bridge the unsuitable area in most cases. Fill the excavated area with large cobbles or pit run gravel. Push the full four foot lift into the depression. Don't do any rolling until the depression is filled to the top. If it is placed and rolled in layers, the gravel will usually continue to roll or pump even at the finished grade level. Naturally, the actual decision to use this method should be made by the inspector, engineer, or property owner. If it does not provide a suitable base, you do not want to be held responsible.

When excavating any road or parking area, try to excavate to the subgrade level before removing the unsuitable material. If the unsuitable soil is so soft that the equipment cannot get through, you may have to remove it before the subgrade level is reached. If the areas to be excavated are too soft, too deep or too small for scrapers, a hoe, track loader, or dozer must be used. If the unsuitable areas are large enough for scrapers and there is a waste disposal area close by, open bowl scrapers with a push-cat can handle the job. If the unsuitable material area is long, either paddle wheel or open bowl scrapers with a pushcat can be used. A paddle wheel scraper can't cut deep enough to get a full load in a short area. Open bowl scrapers can be used in small areas, however, because they have the ability to take a deeper cut.

During the dirt moving operation, avoid running equipment over unsuitable soil areas. This slows the equipment and cuts down production. If there is no convenient way around unsuitable soil, it may be more practical to remove it as it is encountered. This will allow the scrapers to haul over firm ground.

If aggregate is being used to fill the unsuitable areas, and if the unsuitable areas are irregular and hard to measure accurately, you can figure one cubic yard of dirt excavated for every two tons of aggregate used. This rule of thumb is accepted by most contractors, inspectors and owners.

Never try to build a road section over an area that is rolling under the weight of equipment. Invariably unsuitable material will cause the finished surface to break up under the load of traffic. Occasionally you will discover small soft areas after the aggregate base has been trimmed and the road or parking area is ready to pave. A plug of up to a six inch depth of asphalt can be used to bridge these small soft areas. Asphalt will bridge even a very soft area if the asphalt is thick enough. But remember that the asphalt plug must be placed well ahead of the paving equipment so that the plug is hard by the time the top surface is put down. Place the plug the day before paving whenever possible.

A few small unsuitable areas may appear when you are trimming subgrade. If they seem to be shallow, the grader operator can roll the unsuitable material across the subgrade far enough to mix it with dry dirt. After it has been mixed it can be bladed back into the same area and rolled.

Occasionally you will find an area of unsuitable material under a layer of firm earth. This earth will usually bridge the unsuitable material except for a few isolated areas where the equipment has broken through. A situation like this can be troublesome for any contractor. Usually loaded scrapers pulling out of the unsuitable area break off more of the bridging layer. The unsuitable area will continue to get larger at each end of the run as the scrapers enter and leave the area, exposing more of the formerly bridged area on each end with every trip they make. This situation can be resolved in either of two ways. If the majority of the unsuitable soil can be removed without breaking through too much of the bridged area, thus doubling the size of the unsuitable area, scrapers or a loader can be used. Remove all unsuitable material except for the edges where the bridged area has collapsed from the equipment weight. Here a hoe should take over. The hoe can remove the remainder of the unsuitable material without additional deterioration of the bridged area from heavier equipment. If the bridged area is thin, scrapers cannot be used to remove isolated unsuitable areas. The weight of the scrapers will cause the thin bridging to collapse, resulting in larger unsuitable areas. In this case, a hoe should be used to remove the isolated unsuitable areas. The hoe should load the unsuitable material into small dump trucks with a weight capacity less than the scrapers. The unsuitable material can be removed without breaking through the thin bridging areas. This reduces the chance of increasing the size of the unsuitable areas before base can be placed. In some cases the ground may be firm enough to hold a track loader even though scrapers are breaking through in the bridged area. If this is the case, use a track loader in place of a hoe to increase production.

When filling the area where the unsuitable soil was removed, be sure the equipment used is light enough to avoid breaking into more of the bridged area around the edges. A small backhoe with a front loader bucket is ideal for this work.

The inspector may direct the contractor to incorporate the unsuitable material removed into the fill area. If this is the case, the scraper operators must dump only thin layers of soil. Spread the unsuitable material in a thin, even layer and then add a layer of good material. Good material will draw moisture from the unsuitable material when the two are mixed. No water truck will be needed. The moisture from the unsuitable material should provide plenty of moisture.

Sometimes the unsuitable material is so wet that you disrupt the filling operation by trying to incorporate it in the fill. In this case you should be compensated for the production loss. If there is more unsuitable material being directed to the fill than can be incorporated without causing the fill to become unsuitable, you should point this out to the authority involved. An inspector or authority that insists on dumping unsuitable material into a fill area does so at his own risk. If the entire fill becomes unsuitable because of the inspector's error, you are not to blame.

During hot and dry weather, shallow areas that are pumping slightly should dry out enough in one day, especially if they have been ripped. They can then be worked the following day.

Earth that is close to being unsuitable will form slight grooves or ruts as the trucks haul aggregate or pavement over it. These ruts must be rolled flat before spreading aggregate or pavement.

Often you will find unsuitable material surrounding water, gas, sewer, telephone, electrical, or drain lines. This is common where old pavement is to be replaced with new. The asphalt or concrete of the existing road tends to trap water and saturate the soil beneath it. Where utility lines are under the existing pavement and the ground below is soft, there is a real danger that the heavy equipment will break the lines. A loaded scraper can compress the soil enough at a depth of five feet to break a water main. This is probably one of the most aggravating excavation problems because in a short time water will flood the area to be excavated, making work impossible for hours or days. You can avoid problems like this by locating all the utility lines before starting the job. Know both where they are and how deep they are. This is doubly important when unsuitable material is involved.

The best piece of equipment for removing unsuitable material above and around utility lines is a backhoe. In some cases a small dozer can be used to push the unsuitable material out to where scrapers can pick it up. It depends upon what utilities are involved and how much weight they can be expected to take without damage. A reinforced drain line can take much more abuse than a vitrified clay sewer line or asbestos water main. A cast iron water main is a sturdy pipe but can be snapped in the center with excessive weight or a sudden jar. Telephone ducts are as fragile as a clay sewer line. A gas main can probably take more weight without damage than any other utility line.

The utility companies should be notified before the excavation is started. They will be very helpful in locating the lines and service connections. Some utility companies will do the locating for you while others will supply you with a utility plan of the construction area.

If the unsuitable material must be removed below the utility lines or very close to the top, a man on the ground with a shovel and prod rod should direct the hoe operator. Use the prod ahead of the work to keep the hoe operator from damaging any service lines that are attached to the main line. Removing unsuitable material around utility lines is a slow process and should not be rushed. It is an expensive way to excavate. Still, it is much cheaper to take the time to work around the utilities carefully than to damage them. The cost of repairing a telephone cable or gas main is high. Cutting a gas main or an electrical line is dangerous and carelessness may result in loss of life.

Most utility companies will demand that one foot of sand backfill be placed over their lines before backfilling with excavation material. In cases where the line is extremely shallow, the utility company may either lower the line or pour a concrete cap over the line to protect it. You should be reimbursed for any sand used and should not be charged for any lowering of the lines.

After all the unsuitable material has been removed, the inspector will usually decide on the kind of fill material to be used to bring the road back up to the subgrade level. In most cases it will either be dirt, cobbles, pit run gravel, or aggregate sub-base material. After the unsuitable material has been replaced, the area should be trimmed to finished subgrade.

9

Compaction

This chapter describes good compaction practice and shows how compaction tests are made. It also highlights some problems that are common when your work requires compaction testing. There are many types of compaction testing equipment that give accurate readings. There are also many opinions on the type of equipment that is best to achieve good compaction. One superintendent might prefer a cat tractor and sheepsfoot roller on his dirt spread jobs. Another will have nothing but a self-propelled mesh type roller. Most contractors would agree, however, that good compaction is the result of controlling the amount of compaction effort and water on each layer before it is covered with another layer of soil.

Good compaction requires more experience than any other type of excavation work. Compaction is difficult because there are many types of soil and some types require very specialized techniques. For example, sandy soil needs much more water than a heavy clay before it reaches the maximum density or compaction. Be aware of the differences in various soil types and what each type needs when it is compacted. After a few years of experience in compacting different types of soil you will be able to look at a particular soil and know whether it has enough water to compact well. One quick test is to grab a handful of soil and squeeze it. Soil that crumbles when you open your hand is too dry. If it holds solid, it should be good. If you can squeeze moisture out of the soil it is too wet.

The amount of compaction is a measure of the density of a soil. The more dense the earth, the greater the load it will support. Most roads and buildings are designed with the assumption that the soil has a certain density or load bearing capacity. The job specifications usually require the excavation contractor to compact the soil to the density assumed by the designers. Soil tests confirm that the soil has been compacted adequately and that it will support the road or building that is planned.

The two most common types of compaction tests are known as the *sand cone* and *nuclear* tests. The sand cone test is the oldest method of testing. It has largely been replaced by the nuclear test method. Most testing firms say the sand cone method is the more accurate of the two. But the nuclear test is much faster and much more common.

The sand cone test is made by digging a round hole with a volume of 1/10 of a cubic foot. The dirt from the hole is weighed. A cylinder of sand with a known weight is then emptied into the hole until the hole is full. The remaining sand is weighed to determine the amount of sand in the hole. The precise volume of the hole is thus known. The soil removed from the hole is then taken to a soil testing

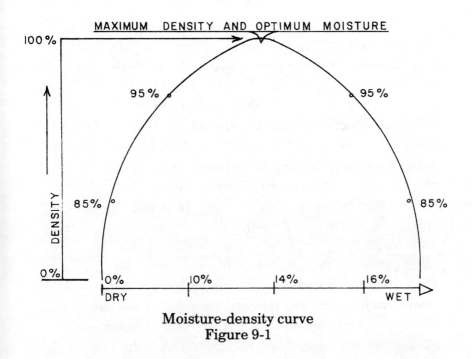

Moisture-density curve
Figure 9-1

laboratory. There it is dried and weighed again to determine how much of its weight was water. An additional sample of soil from the same spot is taken to the lab so a moisture density curve can be plotted on the soil.

A moisture density curve is constructed as follows: Known amounts of water are added to samples of dry soil. The soil is then placed in a tube, tamped a given number of times, and then weighed. Adding water to the dry soil and compacting it will increase the density until it reaches optimum density at a certain moisture. The best moisture level is the volume of water the soil sample contains at the point when it reaches maximum density. Each time a sample is pounded under a given moisture content a certain soil density will result. The percent of moisture and density are plotted as a point on a graph as shown in Figure 9-1. After several samples have been pounded and the points plotted, a line can be drawn connecting the points. This creates a moisture density curve as shown in Figure 9-1.

Water acts as a lubricant and helps the particles of soil slide into place. If too much water is added, the particles of soil tend to float, lowering the soil density. On the other hand, if the soil is too dry, the particles will not slide into the small voids. Thus the density will be lower.

The highest point on the curve is called the point of *maximum density and optimum moisture*. That point is considered 100 percent or optimum compaction. You may see it called 100% A.S.S.H.O. If the lab sample maximum weight is 150 pounds at maximum density and the field sample weight was 145 pounds, then the test result would show 96 percent compaction (150 pounds divided into the 145 pounds field weight and rounded down to the next whole number).

There are two types of nuclear testers. With the newer type you simply set the box on a smooth surface and read the instrument. The older type requires that you drill a small ½ inch hole about one foot deep and then stick a test rod attached to the box into the hole. (See Figure 9-2). Each device works the same way. Nuclear impulses are sent into the soil or aggregate base. A gauge on the box records the impulses reflected by the soil and returned to the machine. The better the soil is compacted, the fewer impulses received and the lower the reading on the gauge. In tightly compressed soil fewer nuclear pulses return to be counted. For example, if the soil is compacted to 95 percent, the reading might be 22,000. If the soil is compacted to 90 percent, you might get a reading of 26,000. Lab personnel

This is an older model nuclear density tester. Notice that a hole has been punched in the ground to get the reading which shows up on a small black screen. Newer nuclear testers give density and moisture readings without inserting a test rod and without any computations.

Nuclear density tester

Figure 9-2

know from the reading given what the density of the material is. They have a chart that gives the particular characteristics of that nuclear tester. This chart is based on the readings the tester gives when placed on a block of material with a known density such as concrete. The nuclear tester is checked occasionally to be sure it is operating properly.

A sample of soil at each nuclear test location is taken to the lab and mixed together. From this soil a moisture-density curve is made using the same method as the sand cone test curve. From this curve the lab establishes the weight of that soil at its maximum density. Once the maximum density weight is known, the lab can determine what the nuclear gauge reading should be at the density required by the specifications.

The correct amount of water must be in the soil or it will not pass compaction testing, regardless of how much you roll it. If soil fails the test because it is too wet, it can be rolled again after it dries and will probably then pass. If the soil did not pass because it was too dry, the only thing you can do is rip it up again, add more water, and re-roll it.

Usually embankment fills must be compacted to 90 percent. In most cases 90 percent is not hard to get. On a large job a greater depth of soil may be put down at one time than the compactor is capable of compacting. Still, the soil may pass the density test if the spread was done correctly. The scrapers hauling in the fill can be used to run over the fill they have dumped the pass before. Thus the scrapers compact the fill as they haul dirt. Even the water truck can do some compacting. Anything running on the fill adds to the compaction. This is why it is much easier to get good compaction in a large fill area than in a small confined area. For example, assume an area 200 feet long and 30 feet wide receives 2500 cubic yards of compacted fill during one shift. An area 800 feet long and 200 feet wide would easily receive 7500 cubic yards of fill using the same compactor but utilizing the scrapers as compactors.

You will have the most problems with density tests during subgrade preparation because usually 95 percent density is required. It is much harder to get 95 percent than 90 percent. To get 95 percent, the soil must be mixed very well. If there are several types of soil in the same fill, as is

WELL GRADED   POORLY GRADED

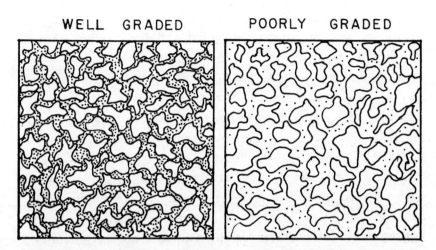

Well graded material compacts more easily than poorly graded material because it is composed of a fairly even distribution of fine, medium and coarse particles. Poorly graded material has voids or air pockets around the larger particles. Every compaction job requires good material gradation, adequate moisture and sufficient compaction effort.

Material gradation
Figure 9-3

usually the case, it takes more working to get the proper compaction. For example, the same fill might have hardpan chips, clay, and sand combined. A few chips of hardpan or a small sand pocket at the point tested can cause the soil to fail. If you are working in a combination of these materials, mix them thoroughly. All the chips of hardpan and sand layers must be mixed with the clay and the hardpan chips must be crushed. See Figure 9-3. Clay and sand together make a bad combination for compacting because clay requires only a small amount of water and sand requires a great deal of water. It is difficult to get the two mixed together without getting too much water in the clay (causing it to pump) or not enough water in the sand.

At 95 percent compaction most soil types show slight movement under the weight of a roller. Some soils show a great deal of movement or actually pump before they reach 95 percent compaction. Trial and error experimentation and a great deal of experience will be needed in many situations. Supervise carefully any time your crews are compacting soil. This is the only way to gain experience.

A vibratory roller is excellent for compacting nearly any aggregate in up to 6 inch layers. In most cases aggregate base material or road rock is easy to compact. Only aggregate material that is too clean or does not have enough sand or rock dust in it will present any problem. Aggregate that has less than six percent passing a number 200 sieve is considered very "clean". Material such as this probably has too much washed rock and not enough crushed rock to bind well. When this is the case, the aggregate must be flooded with water and rolled vigorously to reach 95 percent compaction. In some cases the subgrade or base will be so soft after achieving 95 percent that it must be left idle to dry out before it can be paved. This may not seem logical, but it may be perfectly acceptable from an engineering standpoint.

When trying to compact soil or aggregate material to 95 percent, avoid putting down more than six inches on any one lift. In some cases you will have to compact layers only three inches deep to achieve 95 percent. There are a few soils that can reach 95 percent compaction using eight or ten inch lifts.

A sheepsfoot, grid or mesh type self-propelled roller should roll the subgrade first on any large job. This ensures that there are no dry pockets and that the hardpan chips have been smashed to a size that allows easy mixing.

A pneumatic tired roller is excellent for rolling trimmed
subgrade. It is also a good roller to use on aggregate base. A
tandem or vibratory roller should make the finishing pass
behind it. A rubber tired roller equipped with mats for each
wheel is required as the second roller for asphalt concrete
paving.

Figure 9-4

Following this a multi-tired pneumatic, smooth steel drum
or three legged roller can be used. A rubber tired roller is
always the best for clay type soils. See Figure 9-4. Clay will
not stick to the rubber roller wheels and pull up as it would
with a steel drum roller. For rolling aggregate base material,

a smooth steel drum vibrating roller is the fastest. If the job is large enough for two rollers, a multi-tired pneumatic roller followed by a smooth steel drum vibrating roller is an excellent combination.

Water can be added to dry material and worked in using a sheepsfoot type roller. If a multi-tired pneumatic roller or smooth drum steel roller is used, the moisture needed must be mixed in the soil before rolling because these rollers will seal the surface and very little water will penetrate after the first pass.

The job size will, of course, determine the type of compaction equipment to use and selecting the right equipment is very important. In a large fill job where soil gradation is a problem because you are working with various types of soil, a disc should be used to mix the soil fully. A dozer can be used to pull a sheepsfoot roller and a disc. The sheepsfoot roller mixes the soil slightly. What the sheepsfoot misses, the disc will be sure to mix thoroughly. A disc attachment can also be attached to some self-propelled compactors. Working with a disc on a self-propelled compactor is easier if a hydraulic ram is used to raise and lower the disc.

On a small grade and fill job, it isn't practical to bring in equipment to mix the soil. You can generally get by with only a grader and smooth drum type vibrating roller. The grader can do some mixing as it levels the soil so it can be rolled. This will make a change of equipment unnecessary when you are ready to compact the subgrade and aggregate base. The same two pieces of equipment can be used for both compaction of the existing soil and building up the road cross section.

Once you have excavated a parking lot or road to the

subgrade level, compact the soil at once if it has enough moisture. In many cases a contractor will excavate to the subgrade level and then continue excavating in another area or pull off the job entirely until curbs are poured. When that contractor returns to compact and trim the subgrade, he finds that it has dried out. If this is the case, the area must be ripped up and watered before it can be compacted. The area should have been compacted after the initial excavation while the natural ground moisture was about right for compacting. A vibratory roller would be excellent for getting 95 percent compaction in a case such as this. The area can be left for some time after compaction with little loss of moisture because dense soil loses moisture very slowly. Even though the surface will look dry several days later, neither the compaction nor much moisture will have been lost. The surface can be watered, trimmed, and rolled without any additional compacting. Use this same principle when working fill areas. If the equipment is available, never allow a fill to dry out if you have to return later to compact the top 6 inches to 95 percent.

When several small fills are being made, it is easier to work in several areas at a time. This way the scrapers, water truck, and compactor have more room to operate and will be more productive. Do the work in the following order. The original ground is ripped, watered, and compacted in the first fill area. While the scrapers are dumping a layer of fill across the first fill area, the second fill area is ripped, watered, and compacted. Once the scrapers have spread a layer of fill across the first fill area, they start dumping in the second fill area. The compactor and water truck then move to the first fill area. The layer of fill that was spread by the scrapers is then watered and compacted. When the scrapers have spread a layer of fill across the second fill area, they again haul to the first area which by now has been compacted. As the scrapers move back to the first fill area, the compactor and water truck move to the second fill area to water and compact it. This procedure will continue until each fill has been brought to grade. Work two or more fills in this manner so that all equipment can work at its maximum rate and no waiting is necessary.

10

Excavating
To The
Subgrade Level

This chapter will cover the trimming of subgrade on subdivision streets and highways. It will also explain how and when the job should be staked and the equipment that should be used.

On a subdivision job, the street subgrade is ready to be made up when the underground work and curbs have been finished. The first work is to set the centerline grade hubs. If the curbs are in on both sides of the street, the most common way to set the grade is as follows. Assume the road slopes 2 percent from the center line to the curb in each direction and is 30 feet wide. The surface at the center line would then be .30-foot higher than the lip of the curb (2% x 15 feet). If there are 4 inches of rock and 2 inches of asphalt to be placed, the finished subgrade level at the center line would have to be .20-foot lower than the lip at each curb. When 6 inches (.50 foot) of structure section is added, the finished grade at the centerline will then be .30-foot above the lip of the curbs. This is the 2 percent needed for the crown.

Use three laths, two 3 feet long and one for the center 3.20 feet long. Three men are required with one man at the center line driving hubs. Set the two shorter laths on opposite sides of the street on the curb lip as directly across from each other as is possible without measuring from a station. If the curb stakes are still undisturbed, you can use them as reference points. The longest lath is used on the ground at the center line. Sight with your eye across the tops of the laths. If the center lath is higher than the two side laths, then the grade must be lowered at the center line. If the center lath is lower, then the grade must be raised at the center line. See Figure 10-1. This is a very accurate way to set the grade. If for some reason the lip of the crub on one side is a little too high or low, there will still be an even slope in each direction. Using three laths as described here is called *sweding* and is a fast way of setting grade.

Two other methods of setting center line grades are also used. Some foremen prefer using a string line stretched from curb to curb. A man in the center measures down each time to set the hub. A mark or knot can be made on the string at the center point to eliminate measuring for the road center at each station. If the foreman does not provide an assistant for his grade setter, the grade setter must shoot the grades in with an eye level. This third method of establishing the grade takes one man rather than two or three but is the slowest and least accurate of the three methods. Setting lath with clothes pins may also be considered. See Chapter 3.

After the grade has been set, a grader and paddle wheel scraper can be used to cut the subgrade to the correct level. This is known as *balancing* the subgrade and means that all

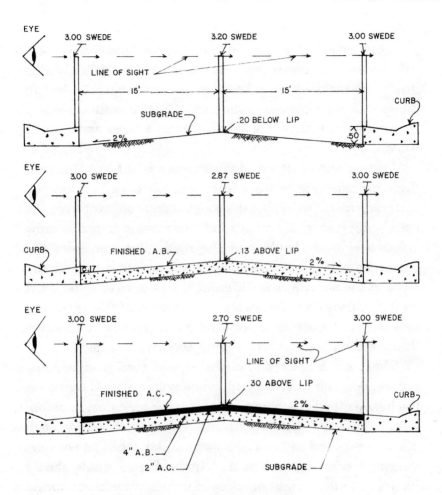

Typical sweding procedure
Figure 10-1

the low and high areas are closely trimmed so they average
.10 foot high. Subgrade soil should be left about .10 foot
high to allow for shrinkage during compaction. After the
subgrade has been balanced, it should be ripped and
watered, mixed to the desired moisture and compacted with

a roller.

Once the subgrade has been compacted and shaped with the grader, it is time to set center line grades again and fine grade the roadbed. In most cases plus or minus .05-foot to .08-foot is an allowable tolerance. This operation usually requires a grader, a small paddle wheel, a roller and a water truck.

Cutting subgrade on highway work is difficult because there are no curbs to work from. Usually on a highway job the grade setter will shoot the rough subgrade level from the cut and fill stakes. He must run three rows of hubs: one row on each side or shoulder of the road, and one down the center line. In some cases five rows are necessary if the slope at the shoulders is different than the road slope, or if an island runs down the center of the road. If the shoulders are not very wide he may elect to carry the road slope through the shoulder area during the rough grading.

There are a variety of shoulder and road designs. See Figure 5-1. The foreman and grade setter should agree on the best method of staking the particular road design before the grade setter starts his staking. Once a method has been agreed upon and the road is staked at the edges and center, the rough trimming can begin. Again, the subgrade should be left .10-foot high to allow for the shrinkage during compaction. Fine grading is always faster when a small trim is needed throughout the roadway. It takes more time when both trims and fills are needed.

When rough trimming has been finished on a highway job and the subgrade has been ripped and compacted, the surveyors will usually re-set the grade before any fine trimming is started. The state, city or county agency decides how the staking will be done. The foreman should consult

with the surveyors if he feels that the staking isn't adequate
for his needs. In most cases, *bank plugs* will be set at each
side of the road. A bank plug is a piece of 2'' x 4'' lumber
driven into the ground and long enough so that it reaches to
approximately 24 inches above ground level. (See Figure
5-6.) On the bank plug the surveyors write the required
information. They drive a nail in each plug at opposite sides
of the street at a given distance above the finished roadway.
If the road has a crown rather than a regular slope in one
direction, the surveyors will set two nails in the bank plug. A
string is stretched from the top nail on one side to the bottom
nail on the other side to set the grade for one-half the road.
The string is then rotated to the other nails to establish the
other half of the road. If the road has a continuous slope in
one direction, it is called a *super* and will only need one nail
in the bank plug on each side. A string is stretched from one
nail to the other and the hubs are set by measuring down
from the string.

In some cases the super in the road will change. One side
may slope 2 percent from east to center line and from center
line to the west edge it may drop 3 percent. In this case each
bank plug will again have two nails so the string can be
rotated to catch the extra 1 percent fall for the remaining
half of the road.

Written on the bank plug will be all the information
necessary for finishing the road. It will have the station
number, the percentage of slope of the road, and a plus or a
minus sign to indicate the direction of the slope from center
line to shoulder. After the bank plugs are set, the grade
setter stretches his string line and sets up his grade hubs by
measuring down from the string.

Some engineering firms do not set bank plugs. Instead,

they set finish subgrade hubs for the contractor. In either case, after the grade hubs have been set, the fine trimming can begin and be finished in the same manner as on a subdivision street.

Control carefully the amount of water used while trimming subgrade. Soil that is too wet or too dry cannot be trimmed properly. The roller operator must wait until the water soaks in before rolling the subgrade. If he does not wait, the roller will pick up the wet soil and leave a rough surface.

A grader operator should have a grade hub set at every 50 feet down the road and at every 20 feet across the road. If the hubs are set at more than 20 feet across the road it is necessary to check the grade just ahead of the grader as it makes the final pass. This is done by using swedes of equal length. Set one lath on each hub and sight across to the center swede. Once the grade has been checked, the cut or fill required can be written on the sub grade soil with paint. If the grader is working with the crew sweding the grade, they can signal a cut or fill to the operator rather than painting it on the subgrade.

A good operator is needed on the paddle wheel scraper for fine trimming. He must be able to pick up the excess dirt without cutting into the trimmed surface. See Figure 10-2. The air pressure in the scraper tires should be checked before any trimming is begun. One tire with as little as 15 pounds less pressure than the others will cause the scraper to lean and dig deeper on the soft tire side as the scraper bowl fills. Even an experienced operator will have trouble when one tire is soft.

Make sure that the top layer of soil is no dryer than the lower layers. Rolling a dry top layer over a wetter bottom

The subgrade has been trimmed to grade and a selfloading scraper is picking up the small windrow left by the grader
Figure 10-2

layer will cause the top layer to crumble and will leave the surface with a scaly appearance.

Don't spend too much time picking out small spots that seem to be a little rough, unless of course the inspector complains. When your grader operator starts working with small areas he will usually create more problems than he solves and certainty wastes time. A lot of time and money can be wasted trimming subgrade if your operator is too careful or is inexperienced.

The operators and the foreman should always keep a close watch on the cutting edge of the equipment. It is costly to repair a worn mole board on a blade or the pan of a scraper. Moreover, a piece of equipment with a badly worn cutting edge is not efficient. For finish work such as subgrade trimming or base rock trimming, a good cutting edge is essential. A badly worn cutting edge is seldom level on the bottom and the ends are nearly always worn. It takes much longer to get the grade within .05-foot with a worn blade. If the cutting edge is bad enough, the blade man will never make the cut properly.

Watch the cutting edge on any paddle wheel scraper that is picking up soil in a trim operation. A scraper with a dull cutting edge or with worn slobber bits will not be able to make a clean pass and will leave more work for the grader operator.

When working a dirt spread that is in hardpan, be sure the ripper teeth are in good condition on the ripper cat, scraper, or grader. Teeth that are sliding in the same grooves each time without ripping are almost sure to be dull.

11

Curb And Sidewalk Grading

This chapter will explain the grading steps to follow for cutting the two most common types of curbs, rolled and vertical. These are usually referred to as type 1 and type 2 curbs. See Figure 11-1.

All the grade stakes must be set before beginning work on the curb. There should be at least a 2 foot offset from each hub to the back of the curb or sidewalk. The stations set by the surveyors should be a maximum of 50 feet apart and all gutter drains and summits should be indicated. A summit is the highest point on the road or street cross section. From the summit, water flows in two directions to low areas where the gutter drains are located. All corners should have a grade stake at the beginning and the end of the radius, and one or more grade stakes between those two points, depending on the size of the radius. A radius point should be staked away from the corner so a tape measure can be used to check the distance to the curb or walk at any point along the curve. Only after the curb has been staked properly

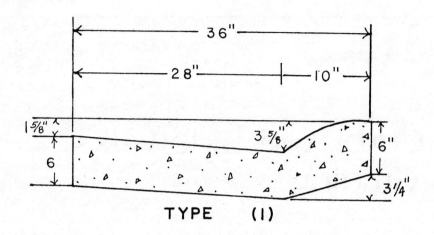

TYPE   (1)

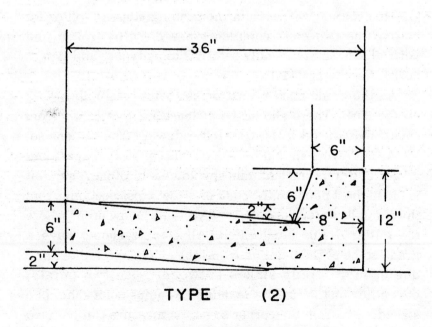

TYPE   (2)

Rolled and vertical curbs
Figure 11-1

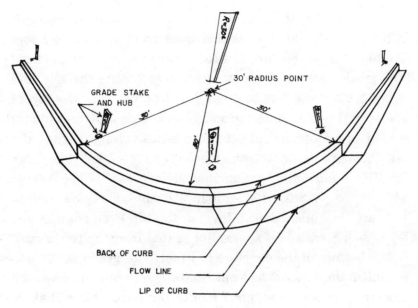

30' RADIUS POINT

GRADE STAKE
AND HUB

BACK OF CURB

FLOW LINE

LIP OF CURB

Radius point staking
Figure 11-2

should the curb grading begin. Figure 11-2 shows a properly staked curb at a corner.

Type 2 curb is the easiest curb to grade. The grade setter sets the curb grade with a straight edge and a hand level. This method is faster than shooting with an eye level. The straight edge should have a one foot vertical extension on one end and a hand level taped close to the opposite end. If the area in front of the grade stake is rough, the grader should make a pass along the front of the stake line to smooth off the area that will be cut. Once a rough pass has been made, the first cut may be started. The grade setter keeps checking the ground level against the grade given for the curb by the surveyors and the grader continues making passes until the soil is at the right level.

Assume that the first stake set by the surveyors reads ②
T.B.C. C-1^{00}. This means that the hub is offset 2 feet and
the top back of the curb is 1 foot below the level of the hub.
The grade setter will measure out 2 feet from the hub and
down 3.05 (three feet and five hundredths) from the top of
the straight edge to the ground level. This 3.05 feet allows
for the one foot vertical extension on the straight edge, the
one foot curb depth, the one foot called for on the stake, plus
.05 (five hundredths) for a slight undercut. C-1^{00} + 1^{00} +
1^{00} + .05 = 3.05. Remember that the surveyors always
indicate the finished curb level at the top back edge.

Check Figure 11-1 again. Notice that from the top of curb
to the bottom of the concrete is 1 foot. The grade setter will
establish the grade 2 feet out from the hub but will have the
grader operator cut within 1 foot of the hub. This will allow
one foot of working room for the forming crew or the
concrete machine. The grade setter holds the end of his ruler
on the spot while he signals the grader operator the amount
of cut or fill remaining at that station. This method must be
repeated at each station until the grade required has been
cut. Once the back of the curb grade has been cut, the grade
setter checks the grade at the front lip of the curb.

Notice that the grade at the front of type 2 curb in Figure
11-1 is 2 inches higher than the back of the curb. The grade
setter measures out 5 feet from the hub and down 2.90 feet.
The 2.90 feet is the result of subtracting .17 (seventeen
hundredths of a foot or 2 inches) from the 3.05 used to cut
the back curb grade. Subtracting .17 from 3.05 leaves 2.88.
Most grade setters will round 2.88 off to 2.90. The front edge
should be very close to grade already if the grader operator
was experienced enough to hold a slight sloping angle while
cutting the back grade. The edge of the blade should pass

over the place where the back of the curb will be when the front grade is being trimmed.

After the curb grade has been trimmed, a windrow of dirt should be bladed up close to the front lip of the curb. The curb crew will use this loose material for re-grading after the forms have been set.

Cutting curb grade for curb forms and for a self-trimming curb machine are nearly the same. But note these exceptions. A three foot offset may be needed on the grade stakes because some machines need more than 2 feet for clearance. The curb grade must be cut .05 foot high to leave the machine a slight trim. Also, no windrow is needed for grading.

Cutting grade for type 1 curb is more difficult because of the small 3¼ inch high slope that must be cut. See Figure 11-1. All measurements are computed from a 2 foot grade stake offset. There is no difference in this respect between vertical curb and rolled curb.

The first cut should be made 1 foot behind the curb. This will usually be 1 foot from the hub. Notice that Figure 11-1 shows the curb 6 inches thick. The grade setter must add .55 foot (6 inches plus a .05 foot undercut) to his straight edge reading plus 1 foot for the vertical extension as is done with the type 2 curb. Adding 1.55 foot to the cut indicated on the stake and subtracting it from a fill will give the level for the top back of the curb only. Once this back grade has been trimmed, the flow line grade still must be cut. First measure out from the hub 2 feet 10 inches. Notice that Figure 11-1 shows a vertical difference of 3¼ inches from the base of the back of the curb to the point below the flow line 10 inches away. The grade setter will convert the 3¼ inches to tenths and hundredths of a foot. Thus he must add .27 foot more to the 1.55 foot he measured down when setting grade for the back of the curb. Now the flow line cut can be made 2 feet 10 inches from the hub. Add 1.80 foot to any cut given and subtract any fill given by the surveyors from 1.80. Adding .27 foot to 1.55 foot equals 1.82 foot. The grade setter rounds the 1.82 off to 1.80.

The grader operator should now make his cut with the blade angled up slightly so the front edge of the curb will be cut to approximately the right grade also. The flow line grade is trimmed by holding the tip of the blade at the flow line and raising or lowering the blade angle to correspond to the angle to the lip of the curb. The drawing of type 1 curb in Figure 11-1 shows all the distance and elevation changes needed for cutting the grade. Notice there is a line extending from the lip of the curb indicating that the lip is 1-5/8 inches lower than top back of the curb. Converting 1-5/8 inches to tenths of a foot would be .13 foot. This added to the 6 inch thickness of curb gives a measurement of .63 foot from the top of the curb to the bottom of the curb at the lip. Adding the .05 foot undercut gives us .68 foot. Round .68 foot to .70. This is the distance used to cut the front lip of the curb. Remember to add the 1 foot for the straight edge vertical extension.

Once the front lip grade is cut, the only excavation remaining is the slope between the top of the curb and the flow line. If the material being trimmed is fine dirt, the curb crew will create this slope by pulling a grade bar with a small tractor. The slope does not need trimming. If the ground is hardpan, the slope from the top of the curb to the flow line must be ripped or cut. It is hard for the grader operator to trim this small slope. The grade stakes obstruct the path of the grader working on the small slope. Many excavation companies have made up curb shoes that can be bolted to the blade. The bottom of the shoe is shaped to match the shape of the curb bottom. The shoe is held on the ground so that the outer edges of the shoe bottom match the back curb and lip grade already trimmed. The shoe will then trim the small slope neatly. The shoe saves some time when cutting only curb grade but will be most useful in getting good production and uniformity when a sidewalk is attached to the curb.

If a sidewalk is attached to the curb, the curb shoe will leave a notch of approximately 2 inches at the point where the sidewalk meets the back of the curb. This notch of two inches is the difference between the 4 inch sidewalk thickness and the 6 inch thickness of the curb where they meet. If a curb shoe is not used, the slope must be cut back a greater distance to undercut for that notch. This will create a small fill at that point for the grade bar when the fine grading is done. When fine grading, a fill is easier to make than a cut. Remember this when cutting curb or sidewalk grade. Always undercut slightly. Any grade that is too high will hamper the hand grading operation. But when a self-grading machine is used, leave the grade .05 foot high so the machine can make a small trim. If hand grading is to be done, leave a windrow of dirt the laborers can use for fine grading. If a sidewalk is attached to the curb, the sidewalk grade is always cut by the grader before the curb grade. Check the grade stakes. If the sidewalk is attached to the curb, the grade given by the surveyors on the grade stakes may be the top back of the sidewalk rather than the curb. If this is the case, check the standard drawings in the specifications to compute the grade changes from the back of the walk to the top of the curb, the flow line, and the lip of the curb.

The only difference between cutting vertical curb grade for a self-grading curb machine and cutting for a formed curb is that you should leave the grade .05 foot high for the curb machine. When cutting sidewalk and type 1 curb grade for the self-grading curb machine, there are two differences. First, the slope between the sidewalk and the curb does not need to be cut. Second, leave the grade .05 foot high rather than undercut it .05 foot. The dirt windrow is not needed for the hand grading. If a self-grading curb machine is to be used and the ground is hardpan, rip the grade ahead of the machine. Ripping and watering the curb grade cut the day before curbing is done will substantially increase production. Any large rocks that may damage the machine should be removed.

12

Widening Rural Roads

This chapter will explain how to widen a road where there are existing driveways, drains, mailboxes, trees, and signs that must be removed or relocated. The type of work this chapter will cover is the common job where an existing road is to be widened six or seven feet on each side. After widening, the new sections have to be brought up to the existing road elevation and the entire new and old road surface has to be overlayed with .17-foot of asphalt.

Road widening is most common in residential areas and working in a residential area is always difficult. The road must be kept free of dust and washed or swept at the end of each shift. More barricades must be used to assure the safety of residents. People who complain about any inconvenience caused them must be treated courteously. On some mornings you may find flooded areas where runoff from lawn sprinklers has accumulated. Good public relations is important. Talking to the residents before starting may save you some time and money. Most people will cooperate if you explain the damage and inconvenience they can cause by allowing surface water to run into the cut and fill areas. Explain to the residents that there may be dust, noise or other problems. Tell them that everything will be done to minimize these problems and ask for their understanding.

The first problem on a job such as this is the existing utilities. Power poles, gas lines, water lines, etc., must be removed or lowered so that they will not interfere with the new construction. After the utilities have been moved or lowered, excavation can begin.

The surveyors will set a line of grade stakes down each side of the road, usually at the right-of-way line. On each stake they will give the information needed to build the road section to the center line.

Once the job has been staked the clearing can begin. Manholes, water valve boxes, or anything in the street that will be paved over must be "tied out" so they can be uncovered and raised later. The best method is to record the station number location of the object being tied out. Then, to be doubly sure of locating it later, measure its location from two substantial objects such as trees, power poles, or fence corners. Take the measurement from each stationary point to the point being tied out and enter the distance under the station number already noted. A drawing should be made. It is a good practice to mark the spot you measured from on the two stationary objects. When it is time to relocate the object tied out, measure from the spot marked on the tree or pole.

All of the home mailboxes should be removed and reset

with dirt in five gallon buckets. The buckets can then be set back out of the way. Many signs must be relocated but also must remain standing during the work. Stop signs, speed limit signs, and other traffic signs should have wooden bases attached so they remain standing but portable. A few sand bags should be placed on the sign stands to keep them from blowing over. Next, grass, brush, fences, and trees should be removed. Any asphalt or concrete that is to be removed must be cut or sawed at the removal line. Next, a backhoe can start removing asphalt or concrete driveways, walks and drainage pipes. After all the clearing has been completed, the dirt excavating can begin.

The back slopes and ditches should be the first excavation work. If there are areas where the ditch bottom is below the reach of the grader, a backhoe must be used.

The main problem in road widening work will be the traffic. All the excess dirt must be pulled onto the existing road shoulder. Here it can be picked up with a small paddle wheel scraper. You need to use part of the road surface for this work. A flagman at each end of the work must control traffic to one direction at a time so that only one lane width of road is needed for traffic at any time. Periodically the traffic can be held up at both ends while the equipment blocks the road.

Driveways should be closed for the shortest time possible. Notify the property owner before obstructing the driveway and give him an estimate of the time his driveway will be closed. He may want to get his car out of the driveway before construction starts. When a driveway culvert is to be placed, be sure that all the material required is on hand before obstructing the driveway. As soon as the ditch is cut to grade, the culvert should be laid and backfilled so the driveway can be re-opened.

Any dirt excavated from the ditch that is free of vegetation should be used in the fill areas. A small dozer pulling a vibratory sheepsfoot roller is a good combination for compacting this or any fill on a narrow shoulder. After the back slopes and ditches have been trimmed, the road structure section can be built up.

The following is the best procedure for excavating, compacting, and trimming a narrow road structure section. Assume the new section is to be .64-foot thick including .30-foot of aggregate base and .34-foot of asphalt. The grade setter should have the grader cut .94 deep at the outer shoulder edge, with the inside edge being held at the top of the existing asphalt. See (1) in Figure 12-1. When this pass has been made, the excess dirt on the street should be picked up. The section is now ready for the second pass.

On the second pass the grader operator holds the tip of his blade right at the inside edge of the existing asphalt. He cuts .37-foot at the pavement edge while holding the shoulder edge of his blade high enough so that loose dirt does not fall in the ditch. See Figure 12-1 (2). Next, the section is ripped, watered, and compacted with a small cat tractor and sheepsfoot roller. After compacting, the grader will make the final passes. The grader operator sets the pavement edge of his blade at the full .47-foot depth. Again he should hold the outside edge of his blade high enough so that he does not lose dirt in the ditch. See Figure 12-1 (3). A smooth drum or rubber tired roller should follow the grader on this pass.

The section is now ready for the final trim pass. On this pass the grader operator will again set his blade against and .47 foot below the existing asphalt. But this time he will hold

(1)

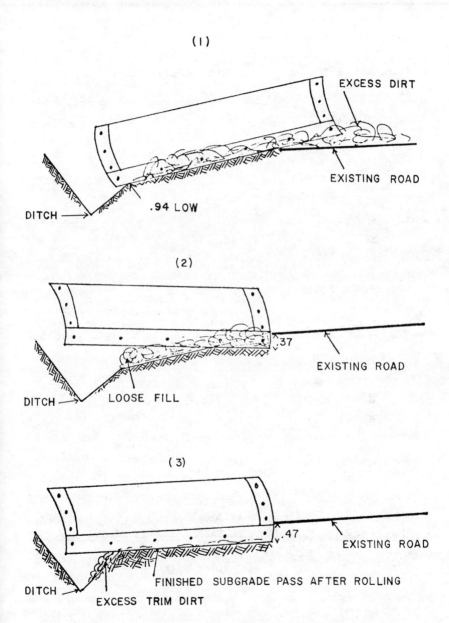

EXCESS DIRT

EXISTING ROAD

DITCH

.94 LOW

(2)

LOOSE FILL

DITCH

.37

EXISTING ROAD

(3)

DITCH

EXCESS TRIM DIRT

.47

FINISHED SUBGRADE PASS AFTER ROLLING

EXISTING ROAD

Subgrade notching
Figure 12-1

(4)

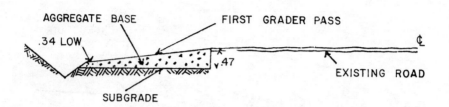

(5)

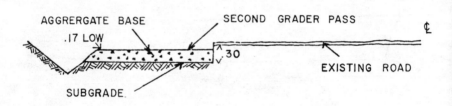

(6)

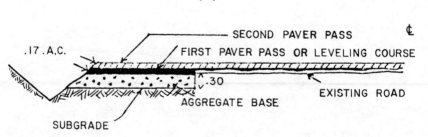

Finished road section
Figure 12-1 (continued)

The small dozer is pulling a vibratory sheepsfoot roller. This works extremely well in narrow areas. The work required a relative compaction of not less than 95%. A tandem roller was used after the grader trimmed the compacted area.
Figure 12-2

the outside edge of the fill on grade, letting any small amount of excess dirt roll off into the ditch. See Figure 12-1 (3). Building the section in this manner should result in very little dirt falling into the ditch.

Aggregate base is placed as shown in (4) and (5) of Figure 12-1. Generally the base is placed in the same

This grader is working in tight quarters. Subgrade is being notched along the existing pavement so the road can be widened. The grader operator is keeping the wheels on the pavement so the rear of the grader does not slide down the slope. Also, work like this is easier if the wheels on the side being cut are on a level surface.

Figure 12-3

manner as the dirt was removed. When placing aggregate in a narrow area, wings should be installed on the grader blade to control the aggregate. See Figure 12-4.

Trucks dumping the aggregate must run with one wheel

The grader is placing aggregate base on a narrow section of
road widening. The wing attachment at the end of the mole
board keeps the aggregate in front of the blade. The small
stake that is about to be covered is a hub. The top of this
stake is at the level the aggregate should reach.

Figure 12-4

in the new section. With the truck bottom dump controlled,
portions of aggregate can be hand dumped right on the edge
of the existing asphalt. The grader tnen spreads the
aggregate, holding the outside or shoulder edge .34-foot

The paving machine is both paving a new five foot section with the first lift and spreading a leveling course out to the quarter crown. Notice the three shades of asphalt concrete. This is caused by the screed dragging gravel at the quarter crown, a smooth pass in the center, and a higher uncompacted edge from the extended hydraulic screed.

Figure 12-5

below the existing asphalt. See Figure 12-1 (4). On the second pass the grader cuts .17-foot into the lift just spread at the existing asphalt edge. The operator should hold the shoulder edge high enough to keep the aggregate from going into the ditch. This leaves the surface level and .17-foot below the existing road. After rolling, the third pass is made holding the edge of the blade .17 below the existing road surface and holding the outer shoulder edge to grade, allowing any small aggregate excess to roll down the slope. This is the same procedure as used with aggregate base shown in (4) and (5) of Figure 12-1.

Next the paving machine can be moved in. The job is paved in two passes. The first pass the paver makes will cover from the shoulder edge to just short of the center line. The asphalt will be laid .17-foot deep at the shoulder edge and tapered to nothing at the center line. See Figures 12-1 (6) and 12-5. This is a leveling course. Now the road is ready to be overlaid from shoulder to shoulder. A final .17-foot of asphalt over the entire road will bring the road section to finished grade. See Figure 12-1 (6).

In warm weather, the asphalt concrete should have three days or more to cool, then, any excess dirt in the ditches should be pulled up on the new pavement and hauled away. The manholes, driveways, and mailboxes are raised, finished or re-set. The street then should be washed and any dikes required should be placed. The street can now be fog sealed and striped.

13

Ripping And Compacting Asphalt Road

Rising construction costs have made it popular to rip and recompact existing asphalt roads to create a stabilized base. This chapter will cover the steps to be followed in doing this work when no grades have been set by surveyors. It may seem easy to do a job with no grades to follow. It is not. With no grades to work to, the crew must establish its own grading plan. Otherwise they will waste time moving material from one side of the road to the other and the cross section will not conform to the road specifications.

The main concern on this type of job is the thickness of the existing road surface. If the job specifications do not supply this information, you will have to dig test holes to determine the exact thickness of the asphalt.

This vibratory roller is completing the pulverizing done by a 64,000 pound compactor. The area being rolled had 4 to 6 inches of asphalt concrete before the ripping and pulverizing began.

Figure 13-1

On this type of a job the surveyors will set roadside stakes and indicate center line distances only. No grades will be set. The job specifications will include a typical road section sheet that shows what work is necessary. The grade setter will set the grades as the work proceeds.

Assume the road section requires that the road be widened two feet on each side and have a 2 percent crown from the center line to the shoulders. The thickness of the asphalt will determine the equipment to be used. If the

This 64,000 pound compactor was used to smash 4 inch to 6 inch asphaltic concrete to a 2½ maximum size. The asphalt was first ripped. A vibratory roller was used for the final pass. The asphalt was pulverized and mixed with the existing aggregate.

Figure 13-2

asphalt is two inches thick or less, a model 12 grader with a rear ripper rack can handle the ripping. It weighs approximately 29,500 pounds and has 135 H.P. The compactor should weigh no less than 38,000 pounds and have a rock tamper or sheepsfoot type drum. If the asphalt is

This class 16 grader was used to rip up a runway where the asphaltic concrete was 4 to 6 inches thick. Most of the ripping was done in second gear with very little strain on the equipment. Three ripper shanks and teeth were used for ripping. Ripper teeth were changed frequently. The shanks lasted only about two shifts.

Figure 13-3

thicker than two inches, the compactor should weigh 60,000 pounds or more. See Figures 13-1 and 13-2.

A model 12 grader can rip asphalt up to four inches thick and still get good production. Use a model 14 grader with a rear ripper rack if the asphalt is four inches thick. The extra weight and horsepower will improve the rate of production and will be easier on the grader. The model 14 grader weighs approximately 40,500 pounds and has 180 H.P. If the asphalt is six inches or more, a model 16 grader or dozer with rippers should be used. See Figure 13-3. A model 16 grader weighs approximately 54,000 pounds and has 250 H.P.

After the asphalt has been ripped, the grader should scrape it into a windrow. This will make crushing the material much easier. The compactor makes a few passes and then the grader operator rolls the asphalt chunks over into another windrow. The compactor again rolls the windrow. This continues until the asphalt chunks are no larger than the required maximum size. Then the material can be spread and rolled to a smooth surface or used as fill for widening shoulders. See Figure 13-4. The ripping and breaking of asphalt should be done on half of the road at a time. This allows traffic to pass during construction.

When the ripped area has been shaped, watered, and compacted, the traffic can be directed over it while the same process is continued on the opposite side. The length ripped down each lane should be equal at the end of the shift. Avoid running traffic over road sections that have an edge of the existing road beside a ripped lane. This might cause a vehicle to lose control.

After the entire length of the job has been ripped, pulverized, compacted, and the shoulder fills are completed, you are ready to begin grading. The best method for balancing the cut and fill areas is fairly simple. Every fifty or one hundred feet the grade setter should shoot the shoulder elevations from one side to the opposite side. If one shoulder edge is .40-foot higher than the other, he adjusts

Asphaltic concrete was pulverized and mixed well into the aggregate base. Here the asphalt and base is being compacted to 95% density with a vibratory roller. The pulverized asphaltic concrete absorbed a large amount of water. A thin layer of new aggregate will be spread over the pulverized area.

Figure 13-4

the grade accordingly. On the side that is .40-foot higher, place a lath indicating a cut of .20-foot. On the side that was .40-foot low, set a lath at the shoulder edge indicating a fill of .20-foot.

Set grades on the entire length of the job in this manner, splitting the difference in grade from one side to the other.

Be sure to check the width from the center line set by the surveyors. The grader operator then makes his cut on the high side to bring the excess material across to make the fill on the low side. This will leave the shoulders level at each side of the road. Now the center line should be checked. Shoot from the shoulder grade to the center line. Compute the percent of slope needed for the crown. Suppose the center line subgrade is .20-foot too high. The following steps should be used to cut it to the finished grade. The grade setter indicates a cut of only .10-foot on the lath at the center line. When the grader makes this .10-foot cut the excess fill is left on the shoulder. Adding .10-foot of fill to the shoulders on both sides of the road will level the road the same as if .20-foot had been trimmed and hauled off. Now the center line will be higher than the shoulder, giving the desired 2 percent slope without using a scraper to balance the grade.

If the center line is low by .20-foot, the grade setter should indicate a .10-foot cut at each shoulder. This would give the center line the fill needed for a 2 percent slope. This is the fastest and simplest method to use when no grades are supplied. The same method is used in the super elevation sections.

Once subgrade has been trimmed and rolled, aggregate grades can be set. The grade setter must set lath at each shoulder edge, measuring up from the subgrade the height of the aggregate desired. Then he places a horizontal line and an arrow pointing to it (a crows foot), indicating the required fill. Setting center line laths would be a waste of time. Traffic and trucks placing the aggregate would knock down the stakes unless the road is very wide.

Keep the aggregate slightly above grade at the road

center and leave both shoulders a little low. The quantity of rock needed every hundred feet should be calculated and that amount should be dumped and spread every hundred feet. Once the aggregate has been placed and rolled the grade setter should run a row of grade hubs down each edge and center line. These are used for the fine grading. He must dig down to the subgrade material at each hub and measure up from the subgrade the appropriate aggregate thickness. This method of controlling the aggregate thickness is used only when there are no grades set by the surveyors.

It is important to leave the aggregate near the center line high because car and truck traffic will cause the rock to separate from the fine material. This is usually called *raveling*. Water the aggregate well, cut the excess from the center line, and spread it to the shoulders which were left low. This will mix the rock and fine material together and leave a smooth, firm base. At the end of each shift, all the aggregate trimmed and finished should be oiled and sanded so the traffic will not ravel it again. The road base is now ready for paving.

After the road has been paved, the aggregate shoulders should be shaped. Then the dikes are placed, the road is fog sealed and then striped.

14

Building Narrow Embankments

Special problems arise when adding a narrow fill section to an existing highway. This is a common problem while widening an existing highway. Suppose the existing highway must be widened 12 feet and must be filled 20 feet to meet the existing road surface. Also, assume that the existing slope is at 2:1 and the new fill must be at 2:1. This means that the fill at the bottom will be 12 feet wide plus 6 feet that must be cut into the existing slope to tie in. This is called *benching* and must continue to the top of the fill. A fill area 12 or 18 feet wide does not allow enough width for the equipment to pass. If a dozer and sheepsfoot roller are on the fill, there will not be enough room for the scrapers to pass.

There are two common ways of building a narrow fill. In the first, the scrapers do not spread soil until all of the

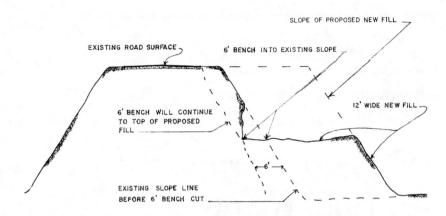

Benching into existing slope
Figure 14-1

scrapers have loaded. When they have all loaded, they move into the fill area together, one behind the other. When the dozer operator doing the spreading and compacting sees them coming, he moves the dozer and sheepsfoot or self-propelled compactor out to the end of the fill and waits until all the scrapers have dumped and passed through. If the fill needs water, the water truck must follow the scrapers over the fill area before the dozer or compactor returns to the fill. The water truck should try to add most of the water needed in the cut area so no passes are needed in the fill area. However, the water truck driver must be careful not to get the cut area so wet that the scrapers lose traction.

This is a slow and expensive operation, but if a lane of the highway cannot be closed, it is the only practical method. See Figure 14-1.

If a lane of the highway can be closed, fill can be dumped on the closed lane and pushed over the side to the dozer and sheepsfoot roller. This is the second way of building up a narrow fill. A model 16 grader can handle about 4,000 cubic

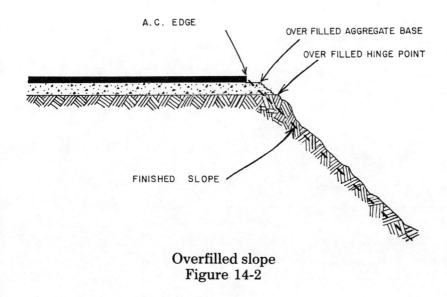

A.C. EDGE

OVER FILLED AGGREGATE BASE

OVER FILLED HINGE POINT

FINISHED SLOPE

Overfilled slope
Figure 14-2

yards a shift in this manner. The water truck with a side spray can then follow the grader, watering the fill when needed. The same operation can be used whether the hauling is done with scrapers or trucks.

If it is a deep fill of 25 to 30 feet and if the dirt is being bladed over the top, a small dozer on the side of the slope may be needed, depending on the material. A heavy clay material will tend to build up at the top of the slope or half way down the slope. You may have to apply an extra shove to get it down to the dozer pulling the sheepsfoot roller at the bottom of the fill.

Operators should keep the outer edge of the fill higher than the inside. This will prevent equipment working on the outer edge from sliding off the fill. Once the fill reaches the subgrade level, it can be worked like any other road section. Be sure the fill is kept 6 inches wider than specified to allow for shrinkage when the slope is compacted. See Figure 14-2.

15

Pavement Removal

Removing an old road and replacing it with a new surface on the same alignment presents problems different from creating a new road alignment. The big difference, other than handling traffic, is that the first layer of material encountered is asphalt rather than top soil.

The most efficient way to remove the asphalt is to use scrapers. If the asphalt is not any thicker than four inches, a scraper with ripper teeth can be used. The scraper should have at least a twenty yard capacity bowl and can be equipped with a paddle wheel. Most foremen believe that asphalt is hard on a paddle wheel scraper and therefore use only open bowled scrapers, a practice that is overly cautious. A paddle wheel scraper with good teeth will pull up the asphalt and stack it neatly in the bowl with no stress to the paddle unit.

Assume that the pavement on a thirty-six foot wide road has to be removed while the existing traffic is allowed to pass. The following steps must be used. Make a cut at each end of the job across the road with a jackhammer, cutting wheel, or hydrahammer. The scrapers start at the right or left of the center line, depending on which side is to be removed first. If the pavement is four inches or more thick, a push dozer tractor may be needed behind the scraper to save tire wear and strain on the transmission, especially if the asphalt is cold. If there are only two inches of asphalt, no pushing is needed. The operator should set the bowl down so the teeth sit just below the bottom of the asphalt. The remainder of the cutting edge, which is higher than the teeth, ride just at the top of the asphalt. This way the asphalt will load easily. The scraper will cut a strip only as wide as the ripper teeth, approximately six feet, and not the full 12 feet of the bowl width.

The next pass the scrapers make will be at the outside edge of the road. After the outside edge and area near the center line have been cut, only an area down the center remains. Remove this in the final pass. Now the grader can level and smooth the road base so that traffic can be diverted to that side while the remaining side of asphalt is removed in the same manner.

The asphalt should not be ripped before the scrapers remove it because ripping makes the asphalt stack up in front of the bowl. The asphalt will slide ahead of the scraper and be harder to load. Ripping the asphalt before the scraper loading starts makes it impossible to use a paddle wheel scraper without damaging the paddle unit. Only an open bowl scraper can be used after the asphalt has been ripped. If possible, all pavement removal and cutting should

The track loader is getting a bucket load of aggregate base and a 2 inch asphaltic concrete section with little effort.
Figure 15-1

be left until midday. The sun warms the asphalt and makes it easier to cut and load.

In most cases where the asphalt is thicker than four inches, it is advisable to remove it with a loader. Either a rubber tire loader or track loader can be used. When using a loader, it is usually advisable to rip the asphalt before loading. This will roughly double the rate of production. A track loader is faster because it has better traction than a rubber tire loader. Good traction is needed to penetrate

Notice that the track loader above is square with the truck while dumping. This is the correct dumping position whether a track or rubber-tired loader is used. The operator has also positioned the bucket so that the material will be dumped in the center of the truck.

Figure 15-2

below the chunks of asphalt when loading. See Figure 15-1. When loading trucks with a loader, it is important to load from a level area and to approach the truck squarely when dumping. The operator must stay far enough from the truck so that the load will be dumped in the center of the bed. See Figure 15-2. On a small job where the cost of moving a ripper cat in would be too high, a track loader can break up the asphalt as it loads.

The route the trucks use from the load area to the dump area should be driven and timed beforehand. Determine the time required for a truck to load, travel to the dump site, unload, and return and then have enough trucks so the loader isn't kept waiting.

16

Lime Treated Base

This chapter explains the three methods used for mixing lime base and outlines the sequence in which the work should be performed. When the soil is heavy clay, a lime treatment is usually called for in the job specifications.

The dirt moving operation proceeds as has been explained previously. The only difference is in the trimming operation. When a lime treatment is to be applied, the subgrade material is always graded as precisely as possible to the correct level before lime is added. This avoids removing large amounts of lime treated base when finish trimming. A little more time spent trimming the subgrade before the lime is added will save time on the final trimming. Usually plus or minus .08-foot is the tolerance allowed for lime treated base. Consequently the subgrade should be trimmed to the same tolerance. Adding the usual 3 to 5 percent of lime does not add enough volume to affect the grade level significantly. The shrinkage caused by compacting will usually offset the volume of the lime added.

When the subgrade has been trimmed the lime spreading operation can begin. The first step is to rip up all of the road surface to be limed. Then one of the following three methods can be used to mix the lime. The first method is to grade the ripped up soil into a windrow, flatten or trench the top, and add the lime over the windrow. The lime and soil in the windrow must be mixed thoroughly by rolling the soil across the road with the blade of a grader several times. This is a slow method but is acceptable on a small job.

The second method is to mix the lime and soil by running a pug mill over the windrow. The grader then spreads the mixture. The problem with these first two procedures is that water must be added to the material while spreading the lime. If all the required water is added before mixing, the material will tend to ball up, making mixing very difficult. Usually the lime cannot be mixed properly when the soil is wet. A pug mill is a rectangular shaped box on wheels with two rows of power driven steel arms that churn the dirt and lime together as it is pulled along straddling the windrow.

The third and fastest way to mix in lime is to use a machine built for the purpose. This machine is similar to a large rototiller. It chews up the soil as it adds water and mixes the lime and soil together. In some cases the mixing can be done in one pass. Even when a lime mixing machine is used, the soil must be ripped before the mixing begins. Then the lime is dumped on the ground at a set rate ahead of the machine. A water truck must move along at the side of the mixing machine to supply water for mixing. Most lime mixing machines have a gauge which indicates to the operator the rate at which the water is added. In most soils 18 percent water must be added to start the action of the lime. In all three operations above, you must have some method of measuring the exact amount of lime being dumped.

A sheepsfoot or self-propelled compactor with steel pads must be used for the first compaction pass. At first the soil may seem too wet, but the lime has a tendency to absorb moisture quickly. A steel drum or rubber tire roller should be used for the final rolling and during the trimming. A rubber tire roller is usually preferred because the material is less likely to stick to the rubber tires.

The lime acts in the soil fairly slowly. After the lime has been mixed, rolled, and the grade shaped, the surface may have to be left until the following morning before trimming begins. The surface must be kept damp at all times to avoid cracking.

The fine trimming is completed in the same manner as you would trim untreated soil. Then the grade should be oiled. Oiling the lime base seals in the moisture, keeps the soil from cracking and allows curing to take place. But don't oil the base if another layer of lime base is to be placed over the existing layer. In that case the lower layer should be kept damp until the second lift is added. A maximum thickness of only 6 inches per lift is usually required by the specifications. In most cases three days must be allowed for curing after oiling before equipment or traffic is allowed on the grade.

Even lime base that has been compacted well and tested soon after trimming may fail the compaction test. If the test results show compaction of 92 percent or better, it is usually safe to go ahead and oil the surface and call for another test after the lime subgrade has had eighteen hours to set. It is not uncommon for lime treated base to pick up 4 percent more on a compaction test after it has another day to cure. Be sure to lay out a few small squares of plywood before oiling so the testing crew will have a few spots without oil where they can take their tests.

If cement is used in place of the lime, the same procedure is used with two exceptions. Not as much water is required, and the fine trimming should be finished and the final rolling completed in 2½ hours. Cement acts much faster than lime. Cement treated base must also be oiled.

17

Aggregate Base

This chapter emphasizes the importance of a controlled dumping procedure and the steps that must be taken to obtain a good finished road base. The speed at which aggregate base is finished is very important. It is also important to have both an experienced grader operator and a good paddle wheel scraper operator for this operation. An inexperienced man making one pass with the paddle wheel scraper can ruin the grade that has been trimmed if he cannot pick up the final trim windrow without cutting into the finished grade.

The surveyors for most agencies will either set bank plugs or grade hubs (blue tops) for trimming the aggregate base. Be sure they are notified in advance to set the grades you need.

A trimming crew should consist of a foreman, grade setter, laborer for hopping guineas, laborer to clean curbs, and operators for a water truck, paddle wheel scraper, roller, and grader. If two graders are doing the trimming, one scraper can still handle the job, but one more roller and one more guinea hopper will be needed.

Laying rock base of any kind should be a relatively easy job. The subgrade has been finished. This gives a finished level to work from. The key to efficient placing of base material is to have a controlled dumping procedure set up. For example, most bottom dump trucks average 25 tons per load and can set the gates to spread the load 100 feet. If 100 tons are needed every 100 feet and each truck dumps 25 tons, have each truck dump a row 100 feet long. Four parallel 100 foot long rows spread evenly across the road will leave the right amount of aggregate. It is important to have even, continuous rows. This insures an even quantity of rock throughout and is much faster for the grader operator spreading the rock. In many cases, and especially on large jobs, a self-propelled spreader must be used rather than a grader. In any case, a good controlled dump is necessary.

Any dumping pattern can be used. It may be desirable to dump one continuous row from the beginning of the road to the end before beginning any of the next three rows. You could also dump four rows side by side in the first 100 feet.

If a self-propelled spreader that runs automatically from a string line or piano wire is used, no grades need to be set. The line is enough. On a small road job where a grader is used, crows feet should be set. Crows feet are lath 12 to 24 inches high with a line marked at the height of the rock grade desired. These crows feet will give the grader operator something to go by for both width and elevation while spreading the base. When there are curbs on both sides, a row of crows feet down the center is all that is needed.

As the rock base is being spread, it is important to get the required water content in the base before it is sealed by the roller. When the rock has a dark grey or brown color and feels damp it is wet enough. After the base has been watered, spread, and rolled, the grade setter or surveyors will set grade hubs for fine trimming.

The grade hubs to which the grader's blade trims are called guineas or blue tops. In most cases the required tolerance for base is plus or minus .05-foot. The grader operator should have a guinea hopper to help while trimming. The guinea hopper cleans the hubs off after the blade passes over them and sets the guard lath back up.

When putting down base with curbs on each side, the excess must be picked up with a paddle wheel scraper. On most roads without curbs the excess rock can be left on the shoulders to be used there after the paving has been completed. It is best to have a small trim to make throughout the road length except for an area at the end. The excess should be placed in a small fill area at the end if it cannot be left on the shoulders. A slight trim is desired because it is faster to trim an area than to stop and make small isolated fills. This will also remove the loose rock left by the trucks hauling aggregate.

Another important point while working rock is to move or blade it as little as possible. Whenever you move rock, the fine materials tend to separate from the larger rock and leave the larger material without the binder needed to hold the rock in place. If this happens, and if it is noticeable, the inspector will require that you re-work the rock to get the fines mixed in again. Rocks that come to the top with no fines to bind them are often called *bones*. The base would be called a bony grade. Water helps fines stay with the rocks while the material is being worked. Water will also help produce a better and tighter grade. The more moist the rock can be kept without making it mushy, the easier it will be to work. The right amount of water will also improve compaction.

On the final trimming the roller operator might need to wait until the rock dries a bit, especially if you are using a steel roller. Saturated rock will stick to the roller drum, leaving a rough surface. If you don't know whether the base is dry enough to roll, make a short pass with the roller. If it is picking up the rock, stop and wait for the surface to dry more. A rubber tire roller will not pick up the wet rock. A drum-type vibratory roller is fast and is excellent for compacting aggregate base. If the drum picks up aggregate, a section of cyclone fencing can be attached over the top half of the drum. This will clean the drum as it rolls.

The most experienced operators should be used for finish grade work because of the close tolerances that are necessary. Remember, if the aggregate base has less than 6 percent passing a 200 sieve, it will probably be necessary to flood the aggregate for it to pass the compaction test.

Most county and state specifications require that the base be oiled if no other base is to be placed on it. If oil is required, it should be done as soon after trimming as possible so little moisture will be lost from the rock base. Do not oil dry aggregate. If the aggregate has dried out, have a water truck make a pass spraying the aggregate before oiling. This will help draw oil into the aggregate. Before shooting oil on any aggregate base, check with the inspector to see how heavy the oil should be sprayed, regardless of what the specifications call for. Coverage between .15 and .20 gallons per square yard of surface is the most commonly used.

18
Working Mud

Working in mud or soft ground that will not support equipment can be aggravating, even for an experienced operator. But experience can help keep an operator out of trouble while working in mud or soft ground. We will explain the best procedure for operating several types of equipment in soft areas.

Track Dozers We start with the dozer because it is the piece of equipment that is the most difficult to pull out of mud if it gets stuck. This is especially true on small jobs where there is only one dozer on the job. When the operator sees that he is about to enter a wet area, he should go slowly. If he feels the front of the dozer start to settle, he should stop and back out. That is the first indication that the ground is too soft. The engine will lug slightly and the front end will start to settle. After backing out, examine the tracks in the soft ground. If they are deep it will be clear that the ground is too soft.

If there is a soft area that must be dozed out to dry or hauled off, the following is the way to go about it. Keep the dozer on firm ground. Start from the edge and work forward slowly. Push the mud ahead of the dozer blade and be sure that the ground below is firm. Do not doze ahead to the point of not being able to get back out. Don't try to move too much material on any one pass. Keep the tracks from slipping and digging in.

Partially stable earth traps many good operators. An operator may drive in and out over relatively firm ground many times as it slowly gets softer. If this happens, the tracks will sink a little more each time until the dozer finally gets *high centered.* In most cases the operator will not notice what is happening because he assumes he can get in and out easily. He is not concerned until it is too late. A track rig is the best equipment to use in mud. But if not used wisely, it can get stuck just as any other piece of equipment.

Scrapers Many times a scraper cannot travel through a muddy area where a dozer can get good traction. In a case where mud is being moved by scrapers, the following is advised. Move into the soft area with the scraper bowl down. Move ahead slowly until the tires start to slip and then stop. Then move the dozer into position to push the scraper. There are two things to keep in mind when the dozer makes contact with the scraper. First, avoid spinning the scraper wheels. Do not try to cut too deeply because this will cause the dozer to spin its tracks and might cause the dozer to get stuck. In either case, the dozer operator will not be able to push on. When the scraper is loaded, apply slightly more power, still being careful not to spin the tires. The dozer should keep pushing until the scraper operator can get enough traction to pull away.

Always watch for soft spots during dumping or loading. If there is a soft area that must eventually be worked through, do not drive right into the center. Drive along the edge first. Move closer to the soft area with each pass. With each pass you can decide whether it is too soft or firm enough to hold the scraper on the next pass. If it is too soft, the compactor can bridge the area with dry fill. If the scraper becomes stuck, the operator should not keep trying to get out. He might be able to work the scraper free if he can see that the wheels are not settling deeper in the mud. This may be the case if the mud is only one or two feet deep and there is a good bottom under the soft material. In this case the operator should swing the nose of the scraper back and forth snaking it out. Another way is to set the bowl down, thus lifting the drive wheels off the ground. The operator can then turn the nose and set the wheels down on a firmer area. Be careful not to make the situation worse. Pulling a well mired scraper out of mud can be difficult.

Compactors Operators should be careful when trying to bridge a fill across a soft area. To get a 24 ton compactor across a muddy area that must be bridged, figure that you need four feet of dry earth pushed out ahead of the machine. The operator should move ahead slowly, being sure that the ground under the compactor is well compacted. A "lugging" engine will be the first indication of trouble. The operator will think he is losing power. Actually the wheels are starting to sink into the fill. When this happens, the operator should back off and push more dry, firm earth over the unstable area, working ahead slowly again until the fill will support the compactor. This same procedure should be continued until the muddy area is bridged. The bridge is complete when a firm pad of dirt is spread completely across the soft area.

The same basic idea can be used in operating any piece of equipment around mud. The main point is to use common sense. Move forward cautiously. Never drive right into the center of a soft area. An operator will find that a little caution will save him many embarrassing moments.

19

Working Rock

Work performed in solid rock is a shovel and truck operation which may require constant blasting. This type of work will not be discussed here. This chapter will cover the type of job where scrapers can be be used (though there is enough rock in the soil to make the excavating difficult) or a solid rock job where the rock can be ripped with a dozer and then hauled with a scraper.

Rock is probably the most difficult material to excavate. An inexperienced operator can do a lot of damage on this type of job. One spin of a tire can cause a blowout. This will ruin most tires and a tire is usually an investment of several thousand dollars. An experienced operator will work methodically in a rocky area. A good dozer operator will study the way the rocks lie and place the dozer blade exactly where he thinks the best leverage may be gained. He will always try to work from a level area he has made for himself.

Trying to doze boulders from a rough and uneven surface is a waste of time and is hard on the equipment. The operator must be thinking ahead at all times. He must pick out voids and then find the correct size rock to fill that void. He cannot just excavate and let the boulders fall where they may. This would be a very inefficient way of doing the work.

Slopes on a rocky job should be cut with a slope bar on a dozer. The operator should run in the lowest gear and make a small cut on each pass. He will leave some boulders sticking out of the slope. Some of these can be left in place and others must be plucked out individually. Usually a two foot plus or minus tolerance is permitted on a rock slope. After a few passes the grade setter can check the larger boulders sticking out of the slope. He should tell the operator which boulders are within tolerance and which should be plucked out. After each slope pass, the operator should level and smooth out the bottom grade the tractor is working from. A rough working surface will result in a rough cut slope.

Loading a scraper in a rocky cut is a delicate procedure. An alert operator is essential. He must be careful not to spin the tires. This will ruin the tires and can cost the operator his job. Apply steady, even power while being pushed. Let the pushing tractor do the work. If the tires start to spin, ease up on the power. If a hard spot is hit and the scraper stops or nearly stops, pull the bowl up just a little until the boulder that was causing the trouble has passed. Then let the bowl down slowly until the push tractor has to strain a little to keep going. The scraper operator can tell when the push tractor's engine is starting to lug down by watching the tractor's exhaust stack. If he sees black smoke start to come out, he should lift the scraper bowl because the push tractor is starting to lug.

The scraper operator will find that in loading rock he must raise and lower the scraper bowl constantly. The cutting edge frequently catches on boulders that will not budge. Do not try to "horse" them out. Let the dozer rip them and then pick them up on the next pass. Occasionally the scraper will load a boulder so large that it will not pass under the bowl when the load is dumped. The bowl cannot be raised high enough. In this case the operator should dump the boulder on the ground, back up slightly to give the scraper a little turning room, and then turn sharply so the bowl passes beside the boulder instead of over it.

Always work slowly and carefully in rock. Do not rush. Working fast cannot make up for the loss of a $5,000 tire.

Running a compactor on the fill produced on a rocky job is also difficult. It is the compactor operator's job to see that the fill does not become so rough that the scrapers cannot move over it. He must either find low areas to push the boulders to or dig a hole for them so they can be covered quickly. Some can be pushed to the outside edge of the fill until they can be covered. But be careful that they are not positioned so that they will be left sticking too far out of the slope. Do not let them slide over the edge of the slope where they cannot be pulled back into the fill. The same cautions should be observed when dozers are working on the fill area. Keep the fill as smooth as possible. Always keep the boulders pushed to the low spots so they can be covered with the smallest effort.

When ripping boulders or rock layers with a dozer, it is usually best to work from north to south or south to north. The cracks in the rocks usually run in that direction. In most cases working north to south makes ripping easier.

Any time you are operating equipment in rocky areas,

think before starting. Make a smooth area to work from as soon as possible. Do not try to overpower the boulders. Doing these two things will increase production.

20

Asphaltic Concrete Paving

This chapter is intended to introduce you to the various methods and equipment used in placing and finishing asphaltic concrete. Self-propelled paving machines are the primary paving tool. Today they are used on nearly all jobs of any size where a specification quality surface is required. Self-propelled paving machines can receive the asphalt in two ways. The first is by end dump trucks dumping mix into the paving machine hopper. The second is by a pickup machine paddling up asphalt which was dumped from bottom dump trucks. The pickup machine works like a self-loading scraper. It picks up the asphalt that has been dumped in a windrow on the ground. The asphalt concrete is then paddled to the top of the pickup machine and flipped into the hopper at the front of the paving machine. See Figure 20-1. From the hopper the mix is carried to the back of the paver by the two conveyor belts and dumped on the

A pickup machine is dumping a windrow of asphalt concrete
in the hopper. The paving machine is starting a new pass. A
pneumatic rubber-tired roller is rolling after the breakdown
roller has finished. Notice one foot along the edge of a
previous pass has not been rolled to make matching
the joint easier
Figure 20-1

ground in front of the screed. See Figure 20-2. Two screed
augers spread the asphalt concrete the entire length of the
screed. The screed strikes off the asphalt concrete at the set
depth. The asphalt concrete passes under the screed and
comes out of the back of the screed smooth and compacted.
See Figure 20-3. If a pickup machine is used, it must be
adjusted periodically so that it scrapes up as much asphalt

Front view of a hopper on an asphalt concrete paving machine. Two conveyors deliver the asphalt concrete to the screed. The side sections of the hopper are hinged and are turned inward by a hydraulic ram. The two rollers in the foreground are placed against the dump truck tires. They push the truck as it delivers hot asphalt concrete to the hopper

Figure 20-2

concrete as possible without cutting into the subgrade.

Two sensors monitor the amount of asphalt which is dumped to the augers from the conveyor. These sensors work independently of each other. If they detect that the augers are becoming overloaded with asphalt, they immediately shut the conveyor off until the augers can distribute the accumulated material. The automatic sensors can usually be turned off so the mix supplied to the augers can be controlled manually. The operator then controls the conveyor delivery rate. Once the augers have distributed the

This paver is starting a new pass. Two men operate the
screed until adjustments have been made. This paver has
one screed extension bolted to each end of the screed. The
finish roller is seen rolling a pass that was put down earlier
Figure 20-3

asphalt the entire length of the screed, the screed both
levels the asphalt at the depth set by the screed man and
compacts the mix with a vibrating tamping motion.

There are two styles of paving machines, track pavers
and rubber tire pavers. The track paver is excellent for
paving on aggregate base, especially if the aggregate is a
little loose. The track paver has much better traction than a
rubber tire paver which will spin the tires occasionally when
the base is loose, and may actually get stuck in the

This is a screed extension for an asphalt concrete paving machine. Several can be bolted on each end of the screed. Auger extensions must be installed to move the asphalt concrete out to the screed extensions if more than one extension is placed on either end

Figure 20-4

aggregate. If the aggregate is firm and if the rubber tire paver does not get overloaded, it will do a fine job. The rubber tire paver is excellent for overlays where it can run on a hard surface.

Most pavers are equipped with a hydraulic screed extension which can be extended out 2 feet when needed. The extension is used when there is a road area that is up to 18 inches wider than the standard width for a short distance. This area can be tapered back to the standard pass width by retracting the extension. The hydraulic screed can be extended to catch the extra width without interrupting the operation. The extension screed leaves the mat higher than the rest of the screed because the extension lacks the vibrating compacting action. The mix is left higher to allow for the extra compression when this area is rolled. See Figure 12-5.

There are many models and sizes of paving machines that with extensions can pave up to 18 feet wide. A few pavers are capable of laying pavement 28 feet wide. Most paving machines are most efficient when paving 10 or 12 feet wide. When the paver is extended past 12 feet, the material must be augered further to reach the extensions at the outer edges. See Figure 20-4. When the paving machine has several extensions, the operator must stop the machine periodically so the augers can catch up with the material requirement of the screed. If a 10 or 12 foot pass is being paved the augers can usually supply an adequate amount of material to the outer edges of the screed. Only when a heavy lift is being placed would the augers fall short of supplying enough asphalt concrete to the ends of the screed if the width is limited to 12 feet.

Using a pickup machine to move hot mix to the hopper will increase the hourly tonnage handled substantially. The big advantage in using a pickup machine is that the paver almost never stops. When using end dump trucks the paver must stop every time the truck is empty so that a full truck can back up to the paver and raise the bed to dump.

The paving machine supplies the power that pushes the truck ahead when end dump trucks are used. The paver also propels the pickup machine which is attached. The pickup machine is only effective on long passes when the paving machine is not required to start a new pass frequently. When a pickup machine is used, the asphalt is dumped on the ground using bottom dump trucks rather than lift bed end dump trucks.

Before paving begins, a string line guide must be set on the ground. The string is set 6 inches off of the center line to prevent the string from being paved over. This will allow the paving machine operator to keep the string line in sight at all times. On a wide road the paving foreman may elect to set a string line at the shoulder edge of the pavement again using a 6 inch offset. Usually no string line is needed when paving a parking lot because curbs serve as a reference line.

Let's assume the roadway being paved is 64 feet wide, 32 feet each side of the crown down the center line. Included in the 64 foot width is an 8 foot shoulder on each side. To pave a road width of this size the following procedure would be used. A string line is set at center line with a 6 inch offset away from the area being paved. Once the string is set, two 12 foot strips will be paved starting from the center line. The two 12 foot passes total 24 feet, leaving only an 8 foot section for shoulder on the first half of the road. The foreman may elect to pave the 8 foot shoulder before the second 12 foot pass is made. It is a good practice to run another string line at the shoulder edge of pavement before paving the 8 foot shoulder pass. The same procedure will be used on the other half of the road with one exception. No string will be needed at the center line because the pavement there is placed flush against the first pass.

If the shoulder grade is the same percentage of slope as the road section, some agencies will allow you to make one 12 foot pass and two 10 foot passes for the 32 feet. Some agencies will allow you to place a crown in the screed and

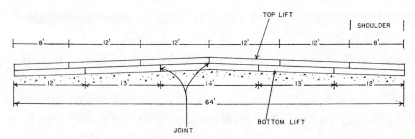

Staggering joints
Figure 20-5

pave a 12 foot pass down the center line, leaving 6 feet in each lane. Then, with the crown taken out of the screed, two 13 foot passes paved on each side of the center line strip would yield the 64 foot total required. If the shoulders have a greater slope than the road section, they must be paved separately. Some agencies insist that pavement joints between passes be at the edge of the lanes where the traffic stripe will be and not in the traveled lane. Read the specifications to know the requirements that must be met.

If two lifts of asphalt are required, the joints of the two lifts should be staggered. If the top joint should crack with age, the bottom lift still remains to hold the surface intact. See Figure 20-5.

Some preparation must be done before a paving machine is ready for paving. Allow time for this. Under average weather conditions the paving screed must be warmed up for 20 minutes. Paving machines have a pump which blows diesel oil spray into the screed tunnel. Here electric fire starters ignite the diesel fuel. While the fire is burning in the screed tunnel and the engine is running, the operator should check the condition of the machine. Any area on the paving machine that comes in contact with asphalt must be sprayed with diesel. This will prevent the asphalt from sticking to cold parts.

If a pickup machine is used, it should be started, checked for worn or loose parts, and sprayed with diesel fuel. The paving machine is equipped with a pump hose and spray nozzle for spraying diesel on any spot where needed. All of the bearings on the pickup machine and the paving machine must be greased and checked for wear. The flight chains that motivate the conveyor on the paving machine must be checked to be sure they have not loosened.

If the mix delivery trucks are scheduled correctly, the paver can continue ahead without stopping. The paver operator should try to adjust the speed of the paver in accord with the number of truck loads of mix available ahead, especially when a pickup machine is being used.

The dump man must be careful to see that the right amount of mix is dumped ahead of the paving machine. The distance each truck load of asphalt is to be spread must be known. Assume that 25 ton bottom dump trucks are used. Each trailer will carry 12½ tons. Assume that we have computed the width and thickness of the asphalt mat being paved. We find that it will take the paving machine 180 feet to use the 25 tons dumped. If you are not familiar with the bottom dump truck spreading characteristics, set the chains on each trailer so the bottom gates will open only 6 inches. Dump the back trailer. Then measure the distance it took to unload. If the distance the one trailer was spread is more than 90 feet, the mix was dumped too light. If the distance was shorter than 90 feet, it was dumped too heavy. If it was dumped too light, have the truck pull out and come around again. The truck should straddle the windrow just dumped. Reset the chains on both trailers to dump a little heavier.

Hand operate the trailer gates to finish dumping the remainder of the mix in the first 90 feet. Once the 90 foot mark has been reached, start dumping the front trailer. Always remember that the complete load must be dumped over no more than 180 feet. If the first trailer dumped too heavy (say it spread only 70 feet), adjust the flow control chains on both trailers a little shorter. Move the truck ahead 20 feet so the second trailer will start dumping at the 90 foot mark. Never dump more than one load in 180 feet. It is much easier to dump a little more asphalt later than to get rid of an excess.

If the bottom dump trucks are going to be hand dumped, which means the dump man will control the gates with a lever, the front trailer should be dumped first. Once the front trailer is empty, the dump man will stand at the spot where dumping was finished until the back trailer gets to him. Then he will continue dumping the second trailer. If the back trailer were dumped first, the truck would have to stop and back up until the front trailer reached the spot where the back trailer ran out of mix. A continuous windrow should be dumped. Dumping the front trailer first is possible only on level ground with a good firm base or when dumping downhill. If the grade is loose or if the truck is dumping on an incline, the truck will lose traction when the front trailer is empty. An experienced truck driver can usually set his chains to dump in the distance asked for by the dump man. See Figure 20-6.

The number of men required with the paver will vary from 3 to 7, including the operator. More men are needed for raking and shoveling if there are cul-de-sacs or tight curves to pave. If a pickup machine is used, another man is needed to handle the dumping. If paving where traffic is a problem,

A good continuous windrow allows the paving machine to
move along at a steady pace
Figure 20-6

two or more men may be needed for flagging traffic.

As paving progresses, the foreman should watch the
temperature, texture and oil content of the asphalt closely. If
the mix does not have enough fine material, it will look rocky
and coarse as it leaves the screed. When the mix is too hot, it
will smoke more than usual and may even look brownish.
The mix will not roll well under the first roller and will tend
to leave a pebbled surface. When the mix is too cool it will
stick to the rubber roller and get lumpy. The amount of oil
can be judged by the shine to the mix. A dull surface means

it needs more oil and a very shiny look means it has too much oil. The asphalt plant should be called if any of these problems occur.

Open graded asphalt is usually spread at a temperature of 200 to 250 degrees. Regular asphalt is usually spread at 250 to 325 degrees. Most job specifications will not allow asphalt to be placed when the air temperature is 40 degrees or less. Open graded asphalt should not be placed when air temperature drops below 60 degrees.

In cool weather the asphalt should be covered with tarps during the trip from the plant to the job. When the weather is extremely hot and traffic must be diverted over the mat just paved before the remainder of the street can be paved, the inspector may require that a water truck spray the asphalt mat to cool it. Traffic can damage a hot mat of asphalt concrete laid on a turn or laid in an area where the traffic must stop and start.

The first strip or mat of asphaltic concrete is put down at the center line and should be laid with the automatic screed height control if the paver is equipped with one. The screed control sensor should be set to drag on the subgrade surface at the center line. The percentage of slope of the road is then dialed into the automatic slope control. On the second pass the screed control sensor drags on the top of the asphalt of the first pass. See Figure 20-7. Any succeeding passes would be done the same way.

The slope percentage should be checked with the indicator at the back of the paver just above the screed. See Figure 20-8. Check the indicator closely to be sure it shows the same percentage of slope as is set on the automatic slope control. The slope indicator is very important when the percentage of slope in the road is changing. For example,

This is an automatic screed control sensor on an asphalt concrete paving machine. An aluminum bar with several sliding feet are used on other models
Figure 20-7

assume the percentage was changing from a -3% to a +3%. As the change is dialed into the automatic slope control, the screed man should watch the slope percentage indicator to make sure the screed is reacting properly.

When starting a pass with the paving machine, place blocks under the screed on each side so that it is off the

In the upper right hand corner is the slope indicator level.
Two elongated nuts are seen in the center of the photograph.
These are turned to create a crown or slight
depression in the mat
Figure 20-8

ground approximately the thickness of the asphalt. Screw
the screed elevation adjustment up or down until a slight
ease in turning is felt in the screed handle on each side.
Then turn the screed adjustment so that most of the weight
is taken off the blocks. This should set the screed so that it
will not drop as the paver pulls ahead and the screed leaves
the blocks. The first few feet traveled on a new pass may
require some quick adjustments. It is good practice to use
two men on the screed for the first few feet. Refer back to
Figure 20-1. After the first few feet only one man is needed.

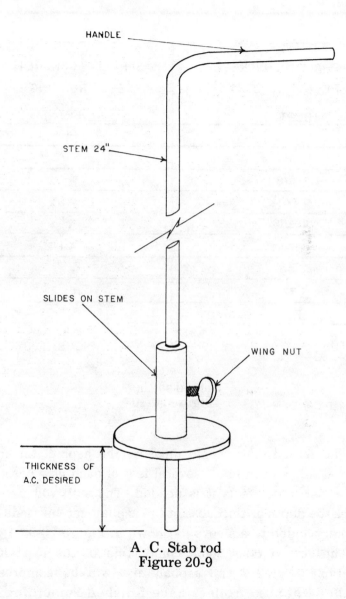

HANDLE

STEM 24"

SLIDES ON STEM

WING NUT

THICKNESS OF
A.C. DESIRED

A. C. Stab rod
Figure 20-9

When you are tying into an existing pavement, a lath should be set under each screed edge to hold the new pavement level up enough to allow for shrinkage that will occur when it is rolled.

ASPHALT CONCRETE SCREED SETTING (DEPTH)	ASPHALT CONCRETE AFTER ROLLING (DEPTH)
1 3/16"	1"
1 3/4"	1 1/2"
2 3/8"	2"
2 7/8"	2 1/2"
3 9/16"	3"
3 11/16"	3 1/2"
4 3/4"	4"
5 3/8"	4 1/2"
6"	5"

DEPTHS MAY VARY SLIGHTLY WITH VARIOUS AGGREGATES

Screed setting
Figure 20-10

The screed man should measure the asphalt thickness several times in the first several feet of each new pass and thereafter as he feels it is needed. The stab rod used to check the depth or thickness of the asphalt can be made of a washer welded to a sleeve as shown in Figure 20-9.

The screed does some compacting of the asphalt as mentioned earlier. The asphalt mat will be compressed further when the rollers have finished smoothing the surface. This shrinkage must be figured into the depth of the asphalt mat as it passes under the screed. You don't want the finished asphalt mat to be thinner than was called for in the specifications. Figure 20-10 shows the depths needed at the screed to achieve the finished depth required after rolling.

There may be times when the paver must straddle a crown in the road or may be required to pave a concave area. The screed on the paving machine can be adjusted for a crown or concave passes. All paving machines are not the same but they all work on the same principle. Above the screed there will be two elongated nuts at the center of the paver. Refer again to Figure 20-8. These nuts can be turned in two directions. The screed is hinged at the center of the paver and will pivot on its hinges as the elongated nuts are turned, one way for crowned mat and the other way for a concave mat. Use a string line stretched across the screed bottom to check the amount of crown or how concave the screed has been set. When using the automatic screed, be sure that the screed arms are screwed to the center before the automatic screed takes over.

The paving machine operator must stop the conveyors each time he comes to the end of a paving pass. He must judge the material and distance so the screed will be empty when the end is reached. It is a good practice to stop the conveyors a little early. More mix can be conveyed to the screed if needed. Any excess material left in the screed at the end of the pass will be left on the ground when the screed is lifted, and must be shoveled back into the machine or spread on the grade where the next pass will cover it.

Paving on and off ramps on highways is usually difficult because the grade changes rapidly from one slope to another. There will usually be a long tapered area to pave and may be a narrow shoulder at a different slope than the traveled lane.

If you are paving a slope that changes from a plus to a

minus grade and if the paving machine has an automatic sensor which senses the grade from a string line or wire, the operator can change the automatic slope control smoothly from one station to the next. To do this the operator must know the percentage of slope at the station where the grade changes before he reaches that station. The grade setter should write the new slope on a 3 x 5 inch card and tack the card on a lath placed before the change station.

Grade changes are more difficult with paving machines not equipped with automatic slope control. Two men should be used to operate the screed. Each man must watch both the thickness of the asphalt being spread on his side and the slope level. The operator should run the paving machine slow enough so that the men on the screed can react to the grade changes. Again, use 3 x 5 inch cards to warn of grade changes so the paving machine operator can let the screed men know in advance of grade changes. If there are shoulders to be paved, they should be paved after the traffic lane has been paved.

When paving tapered areas it is usually best to stop the full pass just short of where the taper begins and then pave the taper first. Once the taper is paved, continue the full pass over part of the taper. This will make a smooth transition and both areas can be rolled together while still hot.

If a pickup machine is used on short radius turns, you can usually paddle enough asphalt from the traffic lane windrow to complete the curve. If more mix is needed to finish the radius, paddle more asphalt from the traffic lane windrow. The same procedure should be used on short tapers. The dump man must be careful with his dumping when a taper or radius is involved. The normal dump pattern might leave material in the way for pulling a taper or radius. The trucks

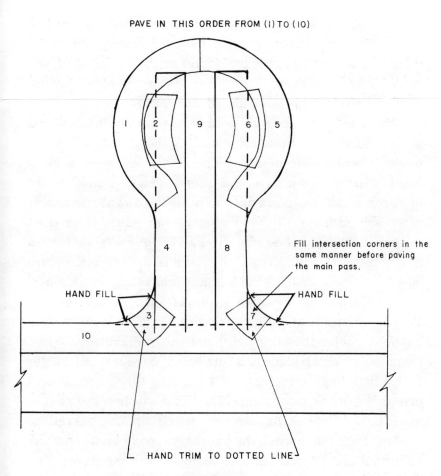

PAVE IN THIS ORDER FROM (1) TO (10)

Fill intersection corners in the same manner before paving the main pass.

HAND FILL

HAND FILL

HAND TRIM TO DOTTED LINE

Cul-de-sac
Figure 20-11

might need to be held up until the taper or radius has been paved before finishing the traffic lane windrow.

The large areas and long passes are paved first. If there are smaller areas where the paving machine must pick up and re-set frequently, the pickup machine should be disconnected. Now the paver cannot use asphalt dumped on the ground in a windrow. End dump trucks must be used to dump mix into the hopper of the paving machine. When the truck has backed up to the paver and lifted the load bed, the truck driver places the truck in neutral and the paving machine pushes the truck ahead as it paves.

When paving subdivision streets, pave cul-de-sacs and short radius curves first before making the main pass. When paving a cul-de-sac or other difficult areas, it is a good practice to use two men on the screed. The first pass should be made at the outside edge, moving in toward the center with each additional pass. The same procedure is used on the opposite side. See Figure 20-11. Intersection curves must be paved before the traffic lane is paved. See Figure 20-11. A good deal of raking and shoveling is required behind the paving machine in areas such as cul-de-sacs. Paving around islands in parking lots can be done with a paving machine though some hand work is involved. When there are several islands in a small area, have a small tractor with a bottom grading blade or back drag fill all the corners just ahead of the paving machine. This will free the paving machine for longer passes and speed up the operation.

There are times when the paver must pave areas so short or curved that a truck cannot maneuver with the paver. The paver must get a full hopper of asphalt concrete, leave the truck, and lay down the short pass or curved area. If more mix is needed, the paver must return to the truck for more mix. If there are many of these areas, it may be faster to dump the asphalt concrete on the ground and have it carried to the hopper with a loader.

The screed man should develop a feel for the particular paving machine he is working on. Each screed has its own characteristics. A screed man should be used on the same paving machine if possible. When paving a second pass parallel to the first pass, be sure that the screed slightly overlaps the first pass. A small gap between the two passes can be filled by raking material into the gap. However, this fill will sink a little after the pavement has been traveled on for a few months. Overlapping passes makes a better joint and saves some hand raking.

If the paving machine becomes overloaded with asphalt because the dump man dumped too heavy, there are two things that can be done. If the paving machine operator recognizes the problem soon enough, he can have the screed man raise the screed slightly until the danger of being overloaded has passed. If the grade tolerance is such that the screed cannot be raised, the paving machine must be stopped and the extra asphalt must be removed to some convenient location. The inspector may allow the paving machine to move ahead and spread a ½ inch thick layer of asphalt well ahead of the paving operation until the excess has been used. When the dump man gets to the area where the ½ inch of mix was spread, he must compensate by dumping less. On some jobs there will be an area available where the excess asphalt concrete can be piled before being hauled off after the job is complete.

Occasionally, the paving machine will have a major breakdown that cannot be repaired before the windrow of asphalt concrete has cooled. If no other paving machine is available to take over, a grader should be used to spread the mix very thin so it can be overlayed when the paving machine has been repaired.

If the paving machine should break down for two or three hours, the mix in the paver will be cold enough to break a flight chain or cause other damage to the paver. Once it has been repaired, raise the screed and pull away from the mat of pavement. If a small amount of mix is in the hopper, shovel it out. The pile of mix left on the ground after raising the screed should be spread out with a shovel. If it has turned to chunks, these should be thrown to the side of the road. When a large amount of mix was in the hopper, it may have stayed warm enough to run through the machine. Try to run the conveyor slowly. If the engine lugs as if it were under heavy load, stop the conveyor. Trying to force mix through the conveyor will usually break a flight chain.

It is important to keep the asphalt truck drivers informed of the paving pattern, especially when several streets are being paved. If there are 20 trucks hauling to the paver, it does not take many confused truck drivers to obstruct the progress of the paving crew. This is especially true of bottom dump trucks because they cannot back up very far without jackknifing. Keep a smooth flow of trucks in and out of the project by making up a map of the paving pattern. Each driver should receive a copy at the asphalt plant. This will eliminate most of the truck control problems and increase production. Figure 20-12 is an example of such a map.

Paving with a spreader box is essentially the same as using a paving machine except for two differences: the spreader box is not self-propelled, and it does not have a vibrating screed. The screed adjustments work essentially the same way but the spreader box has no automatic screed control system and no tunnel to be warmed up. As soon as asphalt concrete slides under the screed on the first pass, stop the spreader box for five minutes and let the asphalt concrete heat the screed. The spreader box is pulled by the truck dumping the mix in it and is attached to the truck by arms that extend into the wheel wells of the truck's back

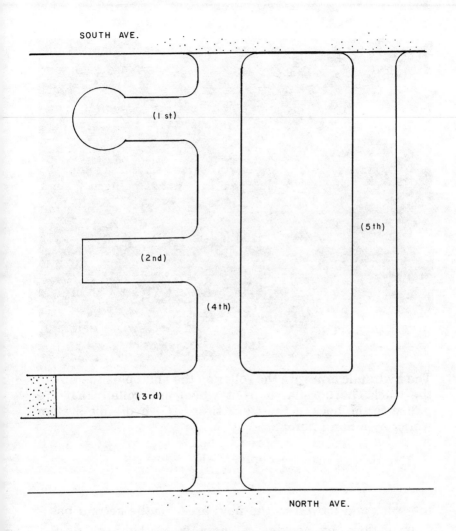

SOUTH AVE.

(1 st)

(2 nd)

(3 rd)

(4 th)

(5 th)

NORTH AVE.

Map for truck haul
Figure 20-12

wheels. See Figure 20-13. These arms have rollers and are pulled tight against the wheel wells with a hydraulic jack type pump. The spreader box rolls on wheels or tracks. When each pass has been finished, the box must be lifted by

The hydraulic arm with the roller on the end holds the box to the truck. The box has a front wheel and rollers that are pulled to the back of the truck tires. The hydraulic system works by a hand pump.

Spreader box
Figure 20-13

the truck and moved to the next area for the second pass. This is done by hooking a chain from the back of the spreader box to the top of the tailgate of the truck. The truck bed is raised when hooking up the chain. When the bed is lowered, the spreader box is lifted off the ground. The hydraulic arms hooked to the truck's wheels stay hooked when lifting and moving the box for the second pass. Most spreader boxes have a hydraulic operated gate that can be shut to hold the mix left in the box. This keeps the mix in the

The screed man makes a screed adjustment as the box is
pulled ahead. The tall handles crank out the screed
extentions on each side as far as 2 feet
Figure 20-14

box when it is lifted. Once the bed of the truck has been
lowered and the box lifted, the truck is free to move to the
point where the next pass begins. Figure 20-14 shows the
box in use and Figure 20-15 is a good view of the rear of a
spreader box.

To start the next pass, the truck bed is raised until the
spreader box is on the ground and there is slack in the
chain. The box can then be unhooked and the gate holding
the asphalt is opened. The spreader box is again ready to go.

Notice the screed adjustment handles, one on each side of
the box. The chain is hooked to the top of the tailgate when
the box must be moved for the next pass. As the bed of the
truck is lowered, the chain will hold the screed off the
ground so only the front wheels will remain on the ground
Figure 20-15

Mix is dumped into the spreader box and paving is resumed.
Chains on the tailgate of the truck must be adjusted so the
tailgate opens only about 15 inches. This will keep the
asphalt from dumping too fast and overfilling the spreader
box. The amount of mix dumped into the box and the speed
the box is being pulled is controlled by hand signals to the
truck driver. Most truck drivers have a tendency to pull the
spreader box too fast. The speed must be kept slow and
even. The screed man needs time to react to the grade
changes and check the thickness of the mat. A steady speed

is required to control mat thickness. If the truck goes too fast, the mat will get thinner; if the truck slows down, the mat will get thicker.

The raker must keep up with the spreader box. If he cannot keep up, the box must be slowed down or a second raker must be used. When a spreader box is used, the asphalt mat should be left approximately 3/16 to 1/4 inch thicker than when using a paving machine. The extra thickness is required because the spreader box does not have a vibrating screed. Mix placed by a spreader box will compact under a roller 3/16 or 1/4 inch more than mix placed by a paving machine.

At the end of each shift, all asphalt must be removed from the box or paving machine. Every surface the asphalt concrete has come in contact with must be sprayed with diesel oil. Give special attention to moving parts. If a pickup machine is used, it too must be cleaned and sprayed.

The number of trucks needed for a paving spread depends on the distance they must travel from the plant to the project. The size of the area being paved and the number of cul-de-sacs, short stub streets, radius, and islands will determine the tonnage that a paving crew is able to put down in a day. A parking lot job with several islands to be paved around where end dump trucks are used might require 125 to 145 tons per hour. If a long section of road is being paved, a pickup machine is used, and bottom dump trucks are hauling the asphalt, the tonnage could range from 325 to 450 tons per hour. Experienced estimators will usually be able to look at a job and determine the daily tonnage. However, it is not uncommon to have to add more trucks or reduce the delivery rate once the paving has begun.

If the superintendent or foreman is in doubt as to how

many tons are needed for a shift, he should seek advice from the estimator or the paving machine operator. Experienced operators usually know how much area they are capable of paving in a day. The driving time between the plant and the job site must be known before any trucks can be scheduled. After estimating the tonnage needed and the travel time from the plant to the job and back again, you should be able to determine the number of trucks that must be used. Always use your high estimate when scheduling trucks. Once the paving begins it is easier to lay off one truck than to add one truck. It is cheaper to have a truck stand by until an adjustment can be made than to have the paving machine and crew waiting for trucks.

When the plant supplying the asphalt is several miles from the job, there may be many trucks on the road at all times. Keep this in mind when coming to the end of the job or at the end of the day. If 15 trucks are hauling asphalt, the truck drivers must be told not to return when only fifteen more loads are needed. If load numbers are marked on each scale ticket, the plant can be called and told to stop loading at a given load number. It is easiest to have the plant control the load cut-off point. But to do this, someone must keep ahead of the situation far enough so that the plant is notified before the last load required leaves the plant. Some asphalt plants hold prepared mix in a hopper. If this is the case, plant personnel must know in advance of the shut off time or final load number so the hopper can be emptied before the shift is over. At the start of the day, check with the plant foreman and ask how much advance notice is needed to clear the hopper.

Once the asphalt is spread, the weight of the roller and number of passes will determine the quality of the surface.

The number of rollers, the order in which they roll, and the number of passes are usually spelled out in the job specifications. When rolling asphalt, always make the first pass on the low side of the asphalt mat. Always roll with the bull wheel first and the tiller wheel trailing. The bull wheel propels the roller and the tiller wheel steers it. If the area being rolled is level, start rolling on the edge away from the side of the next pass. If the paver will soon make the second pass, do not roll the last foot of the second pass side. It is much easier to pave up to an edge that has not been rolled. If the weather is warm, as much as one-half hour can elapse before the edge must be tied into and rolled.

Most asphalt specifications require that three rollers be used: a 12-ton 3 wheel roller or tandem, an 8-ton pneumatic-tired roller, and an 8-ton tandem. All asphalt rollers must be equipped with fiber mats and water to keep the drum or tires wet. Moist surfaces do not collect asphalt as easily as dry surfaces. The number of times the mat must be rolled will also be specified. The initial rolling is done with the 12-ton three wheeler or the 12-ton tandem. This is called the breakdown rolling. The 12-ton roller must keep up with the paving machine, never falling behind. The second rolling is done with the rubber-tired roller and should start when the asphalt mat has cooled slightly, but not cooler than 180 degrees. The third roller is an 8-ton tandem. This is the finish roller. Some vibratory rollers will be accepted as a breakdown roller by some agencies. Once the tires on the rubber-tired roller get hot, the asphalt will not stick to them and water is not needed. The number of rollers required depends on local requirements and the size of the job. Some specifications require only a 12-ton breakdown roller and an 8-ton tandem roller to finish, eliminating the need for the rubber-tired roller. Many small jobs require only an 8-ton tandem for breakdown and finishing. If open graded asphalt is being paved, a tandem roller weighing 10 tons or less is usually required.

An emulsified asphalt (paint binder) must be sprayed on all edges and surfaces where asphalt concrete joins another surface. The only time this tack coat is not needed is when two hot asphalt joints are being joined together. On an overlay job where a new surface is being put on an existing road, an asphalt tank truck, commonly known as a boot truck, can be used to apply the binder. If joints or curbs must be sprayed, an oil pot with a hand sprayer can apply the asphalt emulsions. An oil pot can also be used to spray the tack coat on most small overlays where 200 gallons or less are needed. The amount of tack coat applied is usually determined by the inspector. It may vary from .01 to .10 gallon per square yard of surface. Generally .05 gallon is applied. The asphalt emulsions will be brownish when sprayed on the base or curb. No asphalt should be placed over or against it until it has had time to turn black. If the weather is hot it will turn black quickly. In cool weather it may take 15 minutes. When spraying a tack coat, be careful not to spray it on anything that will not be paved. Use a piece of plywood to collect overspray when spraying against a wall. Sand blasting may be required to remove tack coat that is accidentally sprayed on a rough surface. A rag and diesel will remove asphalt emulsion from a smooth surface.

For small paving areas, patches, and trench paving, an oil pot and small roller are essential. Patching small areas and trench paving takes an experienced crew to keep the patch from becoming rough and uneven. An experienced raker is needed, especially when tapering the edge over

The roller is rolling the feathered edge that was just tied into an existing road surface. The raker is feathering another section that will be rolled next
Figure 20-13

existing pavement. See Figure 20-13. This is called feathering. When feathering is done, three steps must be followed: (1) The existing asphalt must be well primed with an asphalt emulsion, (2) When the asphalt is applied, all larger rocks must be raked out, leaving mainly fine material, (3) The feathered edge must be rolled quickly before it has time to cool. This will leave a smooth edge. The raker must be fast enough to keep the edge from cooling before it can be rolled. A good quality asphalt rake must be used. Otherwise a good job is impossible. This is about the only time the low side of the mat would not be rolled first. If the cold joint being tied into is on the high side of the mat, run a tandem roller on the cold mat, rolling all the asphalt raked over the cold mat plus approximately two feet of the hot mat. Roll as close behind the rakers as possible. Once this is done, the low side pass can be made, working progressively up to the cold edge. As soon as the rakers have more cold edge raked, leave the regular rolling pattern and roll the cold edge again. Hot mix feathered over cold mix must be rolled quickly. Otherwise it will cool and can never be rolled smooth unless more hot mix is placed over it.

The asphalt raker should carry a putty knife with him so he can scrape off asphalt concrete that sticks to the rake. It is important that the rake be kept clean, especially when feathering an edge.

When the asphalt concrete is being placed by hand and a smooth surface free of indentations is needed, the area should be paved in two lifts and no less than a 5-ton tandem roller should be used for rolling. Assuming a 3 or 4 inch depth is needed, place the first lift 2 or 3 inches deep and roll it well. Then tack the edges with asphalt emulsion and pave the last 1 inch. Use this method whenever time permits, regardless of the depth. You will leave a much smoother job if you put down a ½ or 1 inch finish course.

Never use asphalt with aggregate larger than ½ inch when patching or when a smooth feathered edge is needed. For skin patches that must be feathered to match existing areas, use 3/8 inch aggregate mix. When paving small areas by hand, never use any mix larger than ½ inch. Always use ½ inch asphalt when using a spreader box. In many cases ½ inch asphalt concrete will be used for a top lift put down by a paving machine.

A crew that is doing patch work or small paving jobs that involve a lot of hand spreading and tamping of corners should have the following tools: a gas operated plate tamper, hand tamper, spray can, square nose shovels, asphalt rakes, 5-gallon buckets, push broom, picks, rags, oil pot, roller, and diesel for cleaning tools. If the area is very small, a roller and oil pot are not needed. The oil can be spread with a 5-gallon bucket and a brush. A small gas operated plate tamper can be used in place of the roller. A 1-ton vibrating roller is excellent for small areas when a larger roller is not specified. The larger the roller, the smoother the finished product will be. It is very hard to get a good smooth and level surface with a plate tamper.

Edges along curb or walls that cannot be rolled with the roller should be tamped with a plate tamper or a hand tamper before they have a chance to cool. Use a can of water with a spray attachment to moisten the asphalt ahead of the tamper or roller. This will prevent the asphalt from sticking to the tamper plate or roller drum. Most plate tampers are equipped with a small tank for diesel oil or water. The fluid drips onto the plate where the vibrating action sprays it onto the asphalt. Water will keep asphalt from sticking to the plate tamper. Diesel oil must be used on rakes, shovels, and hand tampers. A bucket of diesel oil should be available at all times when paving, regardless of whether paving by hand or machine.

When trench paving, two lifts are advised but not essential unless more than 4 inches is needed. This is the best procedure when paving a trench if 4 inches of asphalt is needed. Dump enough mix in the trench to fill it to the top of the existing asphalt on each side. Tamp it well with a plate tamper or narrow roller that fits the trench width. When it is rolled it should be compressed approximately one inch below the existing asphalt on each side. The existing edges should then be sprayed with an asphalt emulsion. More asphalt is dumped in. This time rake the edges so that only a small amount of fine material extends past the trench edge and leave the asphalt ½ inch higher than the trench edges. Finally, roll this level with the existing surface.

If a large amount of trench paving is to be done, a spreader box can be modified so that it can be used to pave the trench. Plates can be spot welded in the box to narrow the opening to match the trench width. If using a spreader box, apply a tack coat to the edges first. Pave the trench using one lift, followed by an 8-ton roller. If the paving is to be more than 4 inches thick, two lifts must be used, regardless of whether a spreader box is used.

21
Trenching

As in most excavation work, trench excavation requires the right equipment. There are sizes and types of trenching equipment appropriate for any job you have. Your first step is to decide which type of equipment will work best. There are four basic types of trench excavating equipment: hoe, dragline, wheel trencher, and a bucket line trencher.

Generally a wheel trencher is best for depths 6 feet and under where there are no obstacles along the line. A back hoe may be used when the trench is extremely wide or when the trench walls must be sloped. A hoe would also be used if any obstructions such as utility lines, have to be excavated around. On a small job, the high cost of moving a wheel trencher to the site would make using a backhoe more advisable. The advantage of a trencher over a hoe or dragline is the trencher's speed. A dragline must be used if the material being excavated is so loose or soft that a hoe

The long hook that hangs behind the bucket line is the
crumbing shoe. It scrapes the loose dirt
back into the bucket line
Figure 21-1

can't find a firm footing close enough to the edge of the
trench to reach the trench bottom.

When trenching for a water pipe, a string line should be
set to keep the trench in a straight line. No line is necessary
to give the correct depth. If a minimum cover is needed for
the water line, the operator on the trencher can take his
grade from the ground level he is trenching from. The subgrade in most cases has already been cut and depth calcula-
tion should be easy. If a water line is being trenched through
an area that is not cut to the appropriate final grade, a grade

line must be set for depth. A grade string is always needed for line and grade when wheel trenchers are cutting sewer line or drain line trenches.

Wheel trenchers have higher production rates than any other type of trenching equipment. Their only drawback is that most wheel trenchers are limited to a depth of 6 feet and a width of 24 inches. Larger wheel trenchers are made but they are usually not widely available. If a wheel trencher is used, it should be equipped with a crumbing shoe to keep the loose dirt from being left at the bottom behind the wheel. See Figure 21-1. A wheel trencher leaves very little dirt along the top edges of the trench. This makes cleaning the edges easy for the pipe crew. Wheel trenchers have a large wheel at the end of the digging boom and leave more dirt to remove by hand or with a hoe when working close to manholes and utility lines. See Figure 21-2. The bucket line trencher has a smaller digging radius and can move in close to obstacles.

A wheel trencher is more maneuverable than a bucket line trencher and is excellent for short runs like sewer services. A wheel trencher needs a good level surface to work from as do all trenchers. The terrain can roll up and down the length of the trench but must be level at right angles to the trench to keep the trencher from leaning to one side or another. Leaning will cause the trench walls to be slanted. A trench out of plumb is a dangerous place for the pipe crew to work and is hard on the trencher.

Bucket line trenchers are available in various depths and widths. Widths can be varied on most bucket line trenchers. If the trencher available is capable of trenching to the depth needed but is not set up for the width needed, extension brackets can be made to bolt to the sides of every other

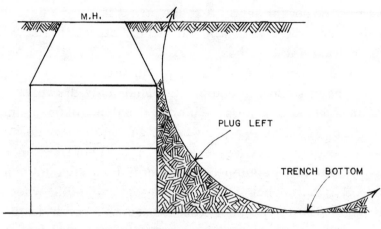

WHEEL TRENCHER

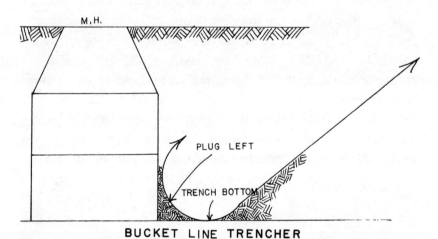

BUCKET LINE TRENCHER

Wheel trencher and bucket line trencher
Figure 21-2

bucket. A two foot wide bucket trencher would then dig a
three foot trench. One or two tooth sets can be welded to the
brackets and teeth set in them. This is a simple job and
should place no unusual strain on the bucket chain. If a four

The arrows point out the rippers attached to the buckets to improve production in hardpan
Figure 21-3

foot wide trench is needed and only a two foot bucket trencher is available, brackets must be made that can be attached to the bucket chain rather than to the buckets. When brackets are attached for extra width the trencher operator will usually need to use a lower gear on the crawl speed and a higher gear for the bucket line and conveyor speed to compensate for the extra dirt being moved.

If the trenching is in hardpan or very hard soil and production is dropping, ripper teeth can be attached to the top of the buckets. See Figure 21-3. These teeth can be welded on every other bucket. They can be staggered so that

Two views of an elaborately designed crumbing skirt. It can
be extended and the pitch of the skirt can be changed. The
chains connect the skirts so they can be raised with the
bucket line when the trencher is being moved

Figure 21-4

Notice that well-designed crumbing skirts keep the edges of the trench virtually free of loose dirt. One man follows behind the bucket line shoveling loose dirt back into the buckets. Two men are busy grading the ditch. The one on the right is checking grade with a level on the grade stick. Notice that the string line is on the operator's side of the trencher. Look closely in the lower right corner to see a one foot step down in the string line on the grade stake
Figure 21-5

two rippers are on one bucket and one ripper is on the next bucket. Setting rippers above the buckets will rip the ground ahead of the bucket teeth and improve production.

A bucket line trencher will leave a great deal of loose material along the top edge of the trench, especially at shallow depths. Skirts can be made to drag alongside the bucket line to push the material back in the trench. See Figure 21-4. The skirt shown in the photographs is very elaborate compared to most. A steel plate bolted to each side of the trencher and attached with a chain to the stinger of the trencher will also work well.

A man in the trench behind the bucket line should throw the loose material at the bottom of the trench back into the buckets. See Figure 21-5. Shoring must be placed in trenches over five feet deep. The shoring must be placed close behind the digging operation so the man crumbing behind the trencher is always protected from a cave-in.

A bucket line trencher can do a good job of trenching starting from a manhole as shown in Figure 21-2. A bucket line trencher leaves a much smaller plug of soil next to obstructions than a wheel trencher. See Figure 21-6. A backhoe is even more efficient around utility lines.

If there are several utility lines to jump, locate them by hand. Then with a shovel dig about 6 inches beyond the line toward the trenching operation and level with the top of the line. Place shredded newspaper or a few shovels full of white lime in the hole on the trencher's side. Mark the spot by painting or staking away from the center line so the trencher operator can see it. The hole can now be refilled. As the bucket line nears the utility line, someone should watch the soil coming out of the trench and be ready to call to the operator. When lime or newspaper shreds are pulled up, the man watching signals the operator. This will indicate to the operator that the bucket line is only 8 to 12 inches away from the utility line. He then should raise the stinger from the trench and pull ahead to just beyond the utility line. The loose dirt over the utility line must be shoveled off so a visual check can now be made.

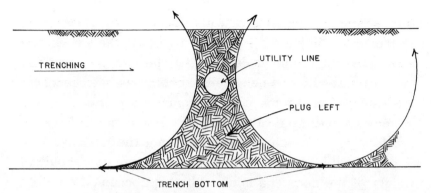

Utility plug left by wheel trencher
Figure 21-6

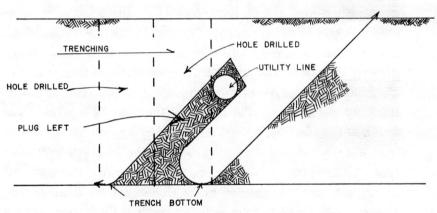

Utility plug left by bucket line trencher
Figure 21-7

Once the top of the bucket line is below the utility line, the trencher operator should dig with the trencher moving toward the utility line, lowering the bucket line until he has reached the depth desired and has trenched to within 6 inches of the line from the back side. See Figure 21-7. Jumping utility lines as described here will leave the smallest plug possible and require the smallest amount of tunneling either by hand or with a hoe.

You can also jump utility lines by setting the trencher in behind the utility line as described earlier and then working away from the line. When the full depth has been reached, the trencher continues up the trench line. This leaves a larger plug of dirt but saves time for the trencher. A backhoe straddling the trench removes the plug the trencher left. This is possible only if the ground is firm enough to hold the weight of the hoe. The hoe can be kept occupied between utility plug work by supplying grading material, digging manholes, backfilling trenches, and setting pipe or manhole barrels.

Another method of jumping utility lines is to remove adjacent soil with a drill rig. The rig should be brought on the job well ahead of the trencher. Determine how much drilling is needed by calculating the length of the plug the trencher would leave around the utility line. The length of the plug depends on the angle of the bucket line and the depth of the trench. When the trencher is working, measure from the end of the stringer to where the buckets leave the trench. This is the distance needed for drilling and determines the number of holes that must be drilled. See Figure 21-7. Once the holes have been drilled they can be backfilled but not jetted or compacted. The trencher then works up to the utility line that has been drilled using the same procedure as if no drilling had been done. Once the trencher resets on the back side of the utility line and digs under the line, the loose dirt from the drilled hole falls into the bucket line. Almost no plug will be left. Usually drilling is considered only in very hard soil. In most soil types the expense would not be justified.

Manholes can be drilled and backfilled ahead of the

trencher in the same manner. When the trencher passes through the soil drilled for a manhole, all the loose dirt falls into the bucket line. This leaves a neat hole excavated for a manhole. The expense of this method must be weighed against the cost of excavating with hand labor or a hoe.

During any trenching operation the ground must be watched closely to prevent cave-ins. If any cracks are noticed, the trenching operation must be stopped until shoring can be placed. Cracks may appear from 1 to 10 feet from the trench edge. When this is the case, the supervisor must decide whether to continue with a trencher or use a hoe so the trench walls can be sloped.

A backhoe can be used any place a trencher can be used. A backhoe does not trench as rapidly but is excellent on short lengths of trench and when many utilities must be jumped. There are many sizes and models of backhoes and most have bucket sizes for nearly any job. Special round bottom, V-shaped, rock and trash buckets are also available. A small backhoe may be the only equipment needed on a small trenching project. The backhoe can trench, supply bedding material, set pipe, and backfill. A backhoe is also a good piece of equipment for digging and setting manholes.

You will get good production using backhoes where utility crossings are a problem if you use two and possibly three backhoes. Assume you figure that your hoe can trench 100 feet every two hours on a particular job and three hoes with front loader buckets are going to be used. Assume the pipe crew is capable of grading and laying 100 feet of pipe an hour when a hoe sets the pipe. Set up a trenching pattern as follows: The first hoe should trench for two hours and then begin setting pipe for the pipe crew. The second hoe trenches for four hours and then begins backfilling the 200

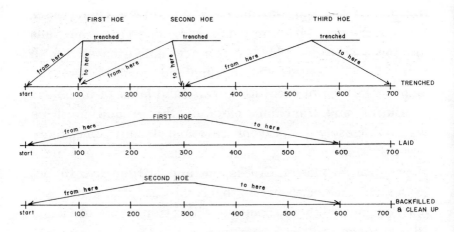

Three hoe trenching pattern
Figure 21-8

feet of pipe that would have been laid by the first hoe. The third hoe would continue to trench the entire 8 hours. Using this plan, by the end of the day 700 feet of ground would be trenched, 600 feet of pipe laid, and all backfilled. See Figure 21-8. There would be 100 feet left for the pipe crew to grade the following morning when the trenching pattern begins the same way. See Figure 21-9. The number of men used for laying pipe depends on the number of hoes and the amount of footage being trenched. This "leap frog" method of trenching works extremely well in firm soil where the backhoes can straddle the trench while digging yet move off the trench without caving in the sides. Straddling a trench with a small backhoe is very simple for an experienced operator.

If the trench sides are too soft for shoring, a hoe can still be used. Instead of vertical trench walls, the trench walls must be sloped a minimum of ¾:1. For every 1 foot the trench rises vertically, it must move in ¾ of a foot

These are two hoes in a three hoe trenching pattern. At the end of the day the pipe laying is within 100 feet of the hoe trenching. The second hoe is swinging a pipe to the trench and the third hoe (not shown) is backfilling. The hoe laying pipe and the hoe backfilling started the day trenching
Figure 21-9

horizontally. This increases the yards of dirt the hoe must move and cuts the production considerably. It also makes using three hoes a little more difficult. When the first hoe gets to the point where the second hoe started, the first hoe cannot straddle the trench if it is being sloped. The first hoe operator must tie his trench into the trench started by the second hoe by digging from the side. Digging at an angle makes it difficult for the operator to keep the spoil dirt away from the trench edge. Usually the hoe must stop trenching to move the spoil pile and begin again.

If the ground is so hard that the hoe is making only slow progress, rippers should be attached to the bucket. Many buckets are designed with slots for ripper shanks. If not, rippers can be attached easily by welding brackets on the bucket.

When trenching around water mains, have a drainage plan worked out in your mind so that if the main is broken, the water can be channeled or pumped away from the work area. Copper or plastic water lines can be crimped to stop the flow. Vice grip pliers work well for crimping plastic service lines. Copper service lines can be bent and hammered shut. If galvanized pipe is cut, a redwood plug can be driven in the pipe to stop the leak.

Always mark utility lines well in advance so that production is not interrupted by a broken line. A broken water main can cause a great deal of flooding and damage in a short time in both the work area and adjacent residential areas.

If a trench crosses utility lines at other than a right angle, the best way to keep hand tunneling to a minimum is to work with the hoe on one side of the trench and then the other, working parallel to the utility lines. This holds true even if there is more than one line to jump in the same area. See Figure 21-10.

A dragline is needed when the trench is deep and the soil is soft. In a situation such as this, a backhoe could not reach back far enough to slope the sides and still reach the material on the bottom of the trench. A dragline is excellent for cleaning irrigation ditches or channels where the reach

Excellent manipulation of the backhoe made virtually no
hand tunneling necessary here. The hoe was able to
eliminate hand tunneling by digging on a 45 degree angle
between the shallow utility lines and again at the
bottom of the picture
Figure 21-10

This is a very deep trench with limited room for a spoil pile. A conveyor belt was set up so the trencher's conveyor dumped into the portable conveyor. The dirt was then conveyed down along the trench and dumped back into the trench behind the pipe laying operation. This allowed the trencher to get rid of the spoil and backfill the trench in one operation
Figure 21-11

required is too great for a backhoe. The major advantage a dragline has over a hoe is its greater reach. If an area to be excavated cannot be reached by a backhoe, a dragline probably will have to be used. A dragline is not as efficient as a hoe because a hydraulically operated hoe has a much faster cycle time than the cable operated dragline. A dragline can be used with an open bucket similar to a hoe bucket, or a clam shell. A few other buckets are available for special projects.

Planning a trenching job can be both a creative and a rewarding task. Many jobs present opportunities for innovative solutions. Note Figure 21-11. Be alert to better, faster, more professional ways to get the job done.

22

Trench Shoring

There are two modern trench shoring systems. Hydraulic shoring is now used in most trench jobs because it is much faster than any other bracing system. Occasionally, however, shoring is done with screw jacks though this requires more men and more time than hydraulic jacks.

To set screw jacks, two men are always needed. A plank must be set on each side of the trench. One man steadies the planks while the screw jacks are placed. The screw jack consists of a 1½ inch section of pipe cut slightly shorter than the width needed to span the trench. One end of the pipe has a foot attached, the other end has a foot with a screw shaft. By turning the wing nut or "ears" on the shaft, pressure is exerted against the planks. See Figure 22-1. One man on a ladder begins by setting the top jack in place. The jack is extended until both ends are against the trench walls. To get the jack tight enough to support the soil, a pipe must be

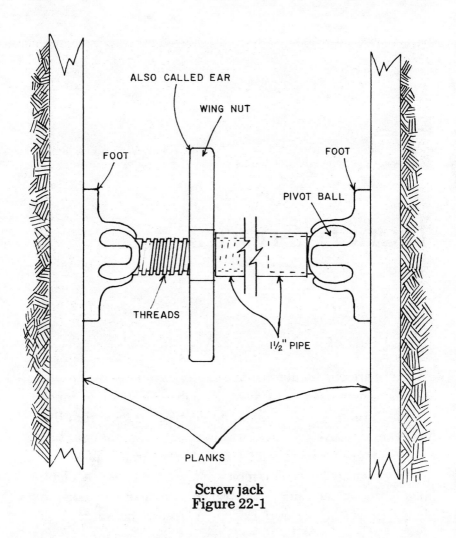

ALSO CALLED EAR

WING NUT

FOOT

FOOT

PIVOT BALL

THREADS

1½" PIPE

PLANKS

Screw jack
Figure 22-1

slipped over the ears and used as a lever to turn the wing
nut. Once the top jack is in place, more jacks are set, from
the trench top to the trench bottom. The number of jacks
placed depends on the trench depth. The maximum distance
allowed horizontally between jacks is usually five feet. When
removing the jacks, work in reverse order, starting at the
bottom and working up.

Three items are needed to set shoring. The hydraulic shore, the pump tank, and the release tool. Notice that the hose is connected to the base of the top jack by a quick coupler. A hose runs from the top jack to the bottom jack so that both cylinders are activated by the same quick coupler. The shore can be collapsed for storing so the aluminum planks are only a few inches apart
Figure 22-2

Hydraulic shoring is both safer and faster than screw jacks. It is safer because no one has to work in the trench. The entire unit can be placed from the top of the trench wall. Various lengths of hydraulic shoring are available. Every shoring section has at least two pump cylinders. See Figure 22-2. Longer lengths have more cylinders. If the shoring is longer than 7 feet, two men will be needed to handle the shore weight.

Use the following procedure when placing hydraulic shoring. The shoring should be set on the ground near the trench in a folded or collapsed position. Locate the hydraulic quick coupler at the base of one of the jacks. Turn the jack cylinder so that the quick coupler is facing upward. Lift the top aluminum plank so the shore is open or unfolded. Once the shore is open, hook the pump hose to the quick coupler which is now readily accessible.

The hydraulic pump tank has a wing nut type valve on top. This valve must be open to connect the quick coupler. After the quick coupler is connected, push down on the aluminum plank so the hydraulic cylinders are completely compressed. With the wing valve open the fluid in the cylinders flows from the jacks to the tank. Now the hydraulic shoring is ready for placing in the trench. The shoring is collapsed or folded again and eased into the trench with the jack hook attached to the shoring handle. The hose, hook, tank, and quick coupler should all be on the same side of the trench. Hold the handle on the plank that will be on the far side of the trench so the shore will not spring open until it is roughly at the right level in the trench. Release the hand held handle and hold the hook side handle. This allows the shoring to spring open. Now steady the shoring with one hand, holding it in place with the hook (the "release tool"). With your other hand, twist the wing nut valve shut. Pump the hydraulic tank arm until the jacks spread enough to push the aluminum planks against the walls of the trench. A firm back pressure should be felt on the tank handle. Now, while still holding pressure on the pump handle, use the release tool to pop the hose coupler free from the jack quick coupler. This completes the installation. Follow this procedure whenever setting hydraulic shoring.

To remove hydraulic shoring, stick the release tool through the handle on the quick coupler side of the shoring. With the hook facing the center of the trench and the cup at the end of the release tool turned inward, place the cup over the quick coupler. Exert pressure against it by pushing out against the handle with the release tool. The fluid will spray out, releasing the pressure on the aluminum planks. Keep pressure against the quick coupler until a substantial opening has been created between the shoring and the trench. Pull the release tool up with a jerk so the hook catches the handle of the shore. The shore can now be pulled from the trench. It will collapse as the handle on the opposite trench side is pulled. If the shoring is long, pull it far enough with the hook to wedge the shoring on an angle against the opposite side of the trench. Then the hook (release tool) can be removed and the shoring pulled the remainder of the way by hand.

The hydraulic shoring used may be too long and heavy for one man to lift on some jobs. If this is the case, a second man can work from a plank laid across the trench. When standing on a plank the second man can reach the far side handle. If the trench is extremely wide, a hook and rope may be needed to pull the far side handle so the shoring will collapse.

Even though the shoring uses hydraulic principles, hydraulic oil is not the fluid used in the jacks. Diesel oil can be used if hydraulic shoring fluid is not available. There are several special fluids made for the hydraulic shoring. Usually you can use a mix of one quart of hydraulic jack fluid and five gallons of water. Hydraulic fluid made for this purpose is much cleaner to use than diesel oil.

Hydraulic shoring is available in various lengths for

Hydraulic shoring 9 feet wide is used against firm earth to protect the workmen from a cave in. Three men should be used to set and release shoring in a trench this wide
Figure 22-3

various trench widths. See Figure 22-3. More than one length of shoring can be used on each point along the trench. For example, if only 6 foot lengths are available and the trench is 12 feet deep, you can use two 6 foot lengths to make a 12 foot shore. This is a good feature of hydraulic shoring. The first 6 foot shore is jacked into place at the top of the trench. Then, using a long release tool, the second shore is placed under the first. Setting hydraulic shoring six feet long is fast because each shore is light and easy to handle.

When setting hydraulic shoring, try to place the shore planks where the trench wall is straight and smooth. If the trench wall is rough or if there is a void behind the jack, pressurizing the shore may bend the aluminum plank. If a smooth area cannot be found, place blocks behind the jack to prevent bending. In sandy soil or gravel too much hydraulic pressure will make the trench wall crumble. When working in soil like this, pump just enough to create a firm pressure against the walls. If the ground is too loose, sheeting must be used. Slide the sheeting in the trench and hold it in place with hydraulic shoring. Check local and federal safety regulations on trench shoring before any shoring is done. These regulations will spell out very clearly how far the shoring must be set apart and how thick the sheeting must be at various trench depths.

There may be areas where ground conditions are so bad that shoring will not hold the trench walls, even when backed with sheeting. If space is limited and it is not possible to slope the bank back from the trench bottom, a shield must be used. There are many types and brands of shields to choose from. See Figure 22-4. Some have adjustable widths and some have fixed widths. A simple shield could consist of two sheets of steel or lighter metal separated by welded steel braces. An open area in the center between the braces allows a pipe to be lowered through the shield to the trench bottom. Most shields must be pulled by a trencher or hoe as the pipe is laid. Some more elaborate shields have a hopper at the front that feeds gravel to the bottom of the shield and a screed that spreads the gravel to the correct grade. Pipe is lowered through the center

Three sizes of trench shoring shields are shown here.
Heights range from 4 feet to 12 feet. Widths are variable.
These shields are pulled along in the trench by the hoe as
the pipe is laid
Figure 22-4

opening and a hopper at the back of the shield backfills
gravel over the pipe as the shield moves ahead. The
excavated material is backfilled right up to the back of the
shield. A hydraulic ram pushing against the backfilled material propels the shield forward. Equipment like this is
usually used by contractors who specialize in large underground projects. Regardless of how elaborate the shoring is,
it is essential that you shore any vertical trench wall over 5
feet deep. Do not place men in a trench deeper than 5 feet

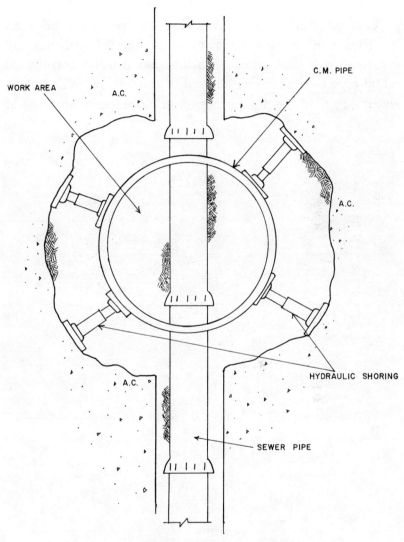

Open center manhole shoring
Figure 22-5

without some type of approved shoring, regardless of how solid the ground appears to be.

Shoring a manhole that must be free of braces so materials can be raised or lowered through it is only slightly more difficult than trench shoring. Set a large corrugated metal pipe or tank casing on end in the hole. Use hydraulic shoring to hold the pipe or casing in place. See Figure 22-5.

23

Laying Water Pipe

This chapter is intended to explain the essential principles of laying water pipe and show you some of the practical tips that are followed by professionals in the field. Most companies that manufacture water pipe offer excellent handbooks that show how to lay the particular pipe they sell. The sales representative who supplies the pipe can get the copies you need.

Good installation practice begins with unloading the pipe. The pipe should be spread along the trench so that you have the required footage between each station. When asbestos cement pipe is being laid, 6 and 8 inch pipe can be laid by hand. All cast iron water line must be set with a crane. The best crane for laying water main is a truck mounted crane. The pallets of pipe are loaded on the truck and unloaded as shown in Figure 23-1.

A full pallet of water pipe was loaded on the bed of the truck and is laid directly from the truck. This is an ideal combination in tight quarters when unloading pipe along the full trench length is not practical
Figure 23-1

One main advantage in laying water pipe as opposed to sewer or drain line is that the trench bottom does not need to be fine graded. The water pipe is laid on two mounds of dirt about 36 inches back from each joint. See Figure 23-2. The biggest concern is laying water main is keeping a constant elevation and line and making each length of pipe line up with the next. If flanged pipe is used, extra effort is

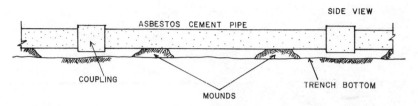

Figure 23-2

required to get the bolt holes lined up. Slip ring joints can be made on a slight angle. However, if the slip ring joint is pushed together on too great an angle, the sliding ring will not seat properly. Forcing the ring may cause the sealing ring to roll out of its seat. This will result in a high volume leak when the water is turned on. The various types of joints are shown in Figure 23-3.

Usually three men are required for laying water main. One man hooks the pipe into place and two do the actual laying. When slip ring pipe is being used, always lubricate the spigot end and not the bell end. Both ends may be lubricated, but never connect pipe with the spigot end unlubricated. The person helping the pipe layer does the lubricating and guides the pipe end into place for the pipe layer. The pipe layer needs a bar to move the pipe into place. It is good practice to protect the pipe by placing a piece of wood between the bar and pipe when pushing the pipe joints together.

In most cases water line trench is from three to four feet deep. This allows the pipe layer and his helper to shovel material from the spoil pile to build the mounds on which the pipe rests. If the material cannot be reached by the pipe layer and his helper, a fourth person is needed to shovel dirt for the mounds.

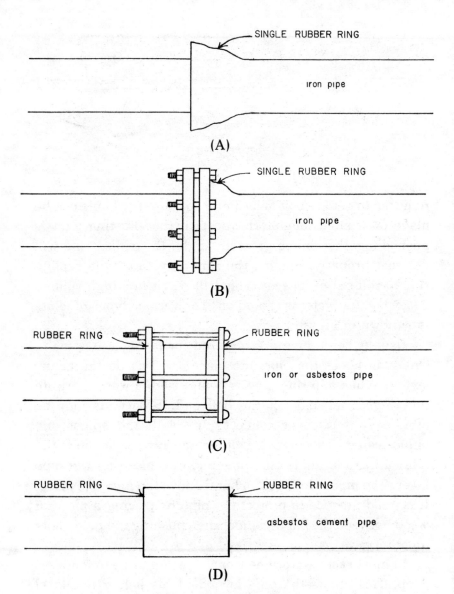

A Bell and spigot
B Mechanical joint
C Mechanical coupling joint
D Slip ring coupling
Figure 23-3

A truck mounted crane holds the cable steady while the pipe
layer's helper lubricates the end of the 6 inch asbestos
cement water line. No sling is used, a quick release pipe
clamp holds the pipe. Asbestos cement pipe is also known
as transite pipe

Figure 23-4

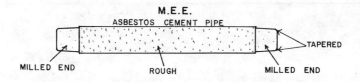

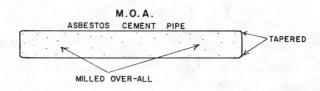

M.E.E. and M.O.A.
Figure 23-5

Laying the main using slip ring joints is a simple operation. See Figure 23-4. Placing valves, tee's, crosses, bends, blow offs, and making water service taps are more involved and time consuming. If asbestos cement pipe is used for water main, all fittings are usually of the slip ring type. Most asbestos cement water line comes in 13 foot lengths. Shorter lengths with milled ends are also available in lengths of 6 feet 6 inches and 3 feet 3 inches. The shorter lengths can be used in combination with the full lengths to avoid cutting pipe when approaching a valve or fitting. Short lengths that are milled on the entire length of the pipe rather than just on the two ends are also available. When the pipe is milled end to end, it is referred to as a *MOA* pipe. MOA stands for *machined over all.* The short lengths that are milled on each end only are referred to as *MEE* sections, *machined each end.* See Figure 23-5. The advantage of the MOA section is that it can be cut to any length and used

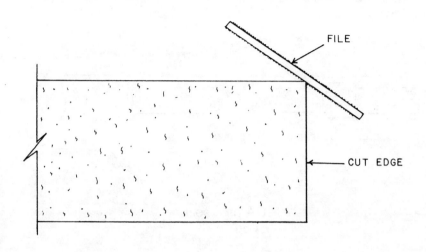

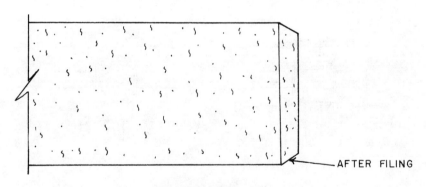

Beveling cut edge
Figure 23-6

without milling the cut end. But remember to file the cut end with a rough file to bevel the end enough so that it slides smoothly past the rubber ring. See Figure 23-6. The MOA section of pipe is also used when tying into another pipe or into a fitting that is stationary. See Figure 23-7.

Never use a compression type cutter on asbestos cement

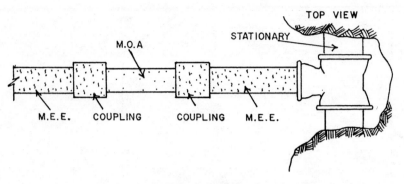

MOA pipe section
Figure 23-7

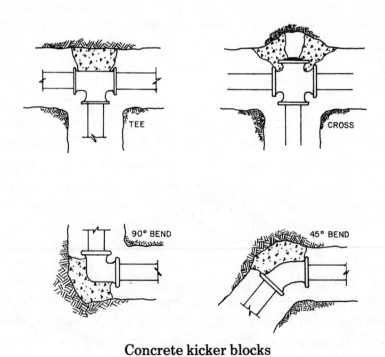

Concrete kicker blocks
Figure 23-8

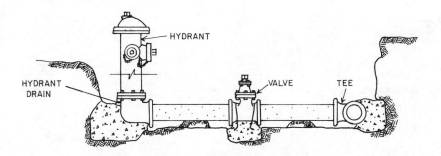

Fire hydrant run and kicker blocks
Figure 23-9

pipe. The correct cutter for asbestos pipe has a knife-like rotating blade or blades. The cutter will cut slightly deeper with each turn until the pipe is finally severed. A portable or truck mounted milling tool is used to mill and bevel the end of MEE pipe once it has been cut. Hand operated milling and cutting tools work as well as power cutters and millers; they just take a little longer.

Usually the job specifications will require that a copper wire be laid alongside or over the asbestos cement pipe so that its precise location can be determined by a pipe locator at a later date. Without the copper wire, a metal detector cannot locate an asbestos line. Job specifications may also require that valves and fittings be set on concrete or redwood blocks. All specifications for laying asbestos cement water line will require that a kicker block be poured at each bend or angle. See Figures 23-8 and 23-9. Asbestos cement pipe can be tied into cast iron or steel pipe by using various fittings designed for this purpose.

There are a variety of ways to make water service taps to asbestos cement water lines. Some asbestos couplings have built-in service taps. You can also drill and thread a hole in

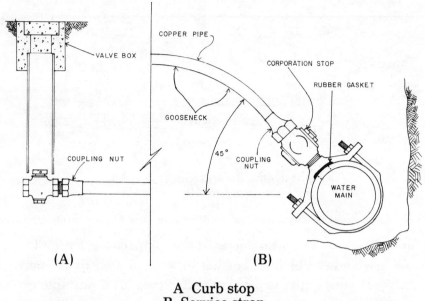

A Curb stop
B Service strap
Figure 23-10

the asbestos main and then screw a corporation stop into the hole. A service clamp is the most common connection. It can be strapped right to the main. Always drill the tap hole after the clamp is on the main. A typical service strap hookup is shown in Figure 23-10.

Any type of service pipe can be connected to the corporation stop. The service line may be plastic, galvanized, or copper, depending on which is specified. Galvanized pipe can be screwed directly to the corporation stop with a coupling. Plastic and copper pipe must be flared. Slide the coupling nut from the corporation stop over the plastic or copper pipe before flaring the end. Once the end has been flared with a flaring tool, attach the service line by sliding the coupling nut down the pipe and screwing it into the corporation stop. All water service lines should have a valve, curb stop, or meter with a valve box and riser at the property line. See Figure 23-10 (A).

If copper is used for the service lines, all joints must be soldered with a copper coupling. Before soldering copper fittings together, be sure all surfaces are clean. A fine emery cloth should be used to sand any area that will be soldered. Then apply the acid solution and solder. Plastic fittings must also be clean and free of moisture before the plastic glue is applied. Keep all galvanized fittings clean and use a liberal amount of joint compound at each joint. When a service clamp is used for a water service tap, use the following procedure. Attach the service clamp firmly to the main, screwing the bolts on each side down evenly until it is tight (55-65 foot pounds with a torsion wrench). Unless the main is very shallow, set the service clamp at a 45 degree angle to the main. After the service clamp is secure, drill the service tap hole and screw the corporation stop onto the service clamp. Asbestos cement pipe can be drilled easily with a masonry drill bit. Once the hole is drilled, the service line can be attached. The service line should have a gooseneck in it as it leaves the main. See Figure 23-10. The gooseneck will absorb any movement or settlement without damaging the service connection in any way. The maximum size of the service tap is limited and varies with the size of the pipe being tapped. Service clamp taps range in size from ½ to 3 inches in diameter. Check with your supplier to get the tap limitations.

When tapping a steel or cast iron main or a live line, a tapping drill must be used rather than the standard drill and masonry drill bit that could be used for dry asbestos cement pipe. There are various types of tapping tools. Some screw to the corporation stop and others are secured directly to the main. When tapping through a corporation stop on a main that is under pressure, follow this procedure. Attach the tapping tool to the corporation stop, making sure the corporation stop is open. The tapping tool should have a ratchet type handle to turn the drill bit. The bit must be ratcheted slowly through the wall of the water main. There should be some tension release felt on the ratchet and possibly some water leakage from the tapping tool when the main is pierced. Be sure the drill bit has gone completely through the wall of the main. Leakage may occur as soon as the bit breaks through the wall. But this does not mean that the tap bit has made the full hole diameter needed for the service tap.

Most tapping tools have a long drill bit. Be careful not to drill completely through the pipe, especially on small diameter mains. Once the tap has been made, the drill bit is backed out of the main and the corporation stop is closed before detaching the tapping tool. Be sure the service lines have been attached and completed and the corporation stops are open before backfilling.

All bends and fittings that require thrust blocks must be poured before backfilling. The job specifications will give the required size of the thrust blocks. Figures 23-8 and 23-9 show the correct positioning of thrust blocks and where they should be poured.

Once the system has been completed, the backfill may begin. The initial backfill must be done carefully because the pipe is sitting on mounds and any sudden jar from backfill material at the joint may fracture the pipe. Carefully tamp backfill material under the pipe to fill any voids. For asbestos cement pipe, sand is an excellent initial backfill material. Be very careful while backfilling. Don't let any large chunks of dirt or rocks lodge against the pipe.

After the trench has been backfilled and compacted or

water jetted, testing can begin. Check the job specifications for test requirements. Some agencies require that all joints and fittings be exposed during a pressure test and that just enough backfill be placed on the pipe to keep it from moving under pressure. If this is the case, backfilling and jetting will be completed after passing the pressure tests. The job specifications will also specify a length of time concrete kicker blocks must cure before applying pressure. Usually 24 hours is required. The pressure test in some cases may consist simply of slowly opening the valve from the live main until the new line is filled. After a few hours the inspector makes a visual check of all the joints and fittings.

A more difficult test requires that extra pressure be pumped into the line. Complete backfilling and compacting is usually required before this type of test begins. Start by slowly opening the valve from the live main to the new main. Hydrants along the system should be cracked open to release air in the line. They must be shut quickly when all air has escaped and the water flows evenly. Enough water should be drained off to release all the air but not enough to lose any chlorination that may be in the line. Once the air is released and the new main is full, the valve feeding the line must be shut down. A pump is needed to add the required pressure. Usually the pressure specified will be 150 to 200 pounds per square inch. The pump draws water from a 55 gallon drum and pumps this water into the system. The discharge hose from the pump to the hydrant must have a valve and a pressure gauge before the coupling attaching it to the hydrant. The valve from the pump to the hydrant is closed when the required pressure is reached. The pump is then shut down. The amount of time the test pressure must be held varies with different agencies. Some agencies test by the pressure lost. Most agencies measure the gallons needed to refill the line after a period of time. This is done by measuring the amount of water drawn from the 55 gallon drum when the pump refills the line to the original pressure. The amount of leakage allowed will vary from one agency to another. The specifications give you the allowable amount.

Once the results of the pressure test are acceptable, the main must be chlorinated. One of two methods can be used. In the first, a specified number of calcium hyprochlorite tablet are placed in each joint of pipe as the pipe is laid. These tablets are fastened to the inside of the pipe with permitex or rubber cement purchased at any auto supply store. Apply permitex or cement to only one side of the tablet, leaving the remainder of the tablet exposed to the water. The job specifications dictate the exact number of tablets needed in each pipe joint. If tablets were placed while laying the line, the line cannot be flushed before testing. Great care must be taken while laying pipe that cannot be flushed to keep dirt and foreign matter out of the pipe. Chlorination testing is done after the pressure test and the main should already be full of water. The valve on the live main is opened slightly to allow water into the main as water is released at the end of the line. A hydrant at the extreme end of the new system should be opened slightly until water runs slowly. After several gallons have run through, the water should be fairly clear. At this point, the agency will test a water sample. If the hyprochlorite tablets have dissolved correctly and care was taken to keep dirt out of the line, you will pass the chlorination test easily. The main must remain full of water for a certain time before the test is made but the time needed for the pressure test usually meets this requirement.

If tablets were not placed in the pipe when it was laid, calcium hyprochlorite must be pumped into the system. With the main full of water and flushed, turn the controlling valve off. A hydrant or service valve close to the controlling valve at the live main end is connected to a pump and reservoir of calcium hyprochlorite dissolved in water. A hydrant or service valve at the far end of the main is opened. Then the liquid calcium hyprochlorite is pumped into the main until it starts coming from the hydrant valve at the far end. You will be able to smell the calcium hyprochlorite as it comes out of the hydrant. Once this occurs, shut everything off. The line is now chlorinated and must sit a certain amount of time before a sample can be taken. The reservoir for the mixture of water and calcium hyprochlorite will usually have to be a water truck because you need enough volume to fill the line. Once you have passed the chlorination test, the line should be flushed well.

24

Laying
Sewer Pipe

The starting point for every sewer laying job is unloading the pipe correctly. Where you unload the pipe depends on how the pipe will be placed, by equipment or by hand. If the equipment used has a swing boom, the pipe should be placed far enough from the trench so that the equipment can drive between the trench and the pipe. If the equipment used has a fixed boom that will only raise and lower loads, the pipe should be unloaded along the trench edge. The pipe should also be placed along the edge of the trench if it is going to be lowered by hand or rope. Never try to save money by hand lifting heavy or long sewer pipe.

If pipe is being unloaded on pallets, it is important to know how many linear feet are on each pallet. The pallets should be spaced so that there is the right footage of pipe between each trench station. See Figure 24-1. The pipe

This 6 inch clay pipe is light enough to be spread by hand. Consequently, it is placed close to the surveyor's lath that mark the trench line. The grade has been smoothed for the trencher and the lateral pipe and Tee's have been stacked to the back for easy access. A precast manhole bottom has been unloaded where it will be set

Figure 24-1

laying operation will be delayed if the men and equipment must go up or down the line for additional pipe.

Be prepared for the pipe when it arrives. Have an unloading pattern already worked out and have the correct equipment available for the unloading. To be sure that you have the correct equipment on the job, you must know the

A truck mounted crane is using a rope sling
to unload clay pipe
Figure 24-2

weight of the pipe or pallets. Have the right slings or hooks available so you can unload the pipe or pallets without damage. See Figures 24-2 and 24-3. The manufacturer can supply the weight information needed and explain how the pipe will be loaded. He can also suggest how to unload the pipe. Many pipe manufacturers will supply the slings or hooks needed for unloading. Talk with the manufacturer before receiving the shipment. A few manufacturers will do the unloading but you still must direct the man unloading the pipe to the correct area.

Once the pipe has been unloaded and any crushed rock or sand required has been dumped in convenient locations,

A large hoe unloads 84 inch concrete pipe with a cable sling.
Notice that the bucket has been removed
Figure 24-3

preparations for laying the pipe can begin. The trench must
be dug, shored, and a grade line set. Refer to the chapters
on trench excavating, shoring trenches, and grade line
setting for detailed information. When the shoring and
grade line have been placed, the hand grading can begin. If
no special bedding material is specified, trench spoil can be
used. An improperly graded trench can cause two serious
problems. It will make sliding the pipe together difficult,
especially if large pipe is being laid. Poor grading will also
interrupt the flow through the pipe. A poorly compacted
trench bottom may cause the pipe to break from settlement

after being backfilled and compacted or jetted. It is important to have experienced men doing the grading and laying.

For grading, one man must be in the trench doing the actual grading while one or two men are on top of the trench. One man shovels material down to the man grading as he needs it. If a top grade line is required, the second man holds the grade rod and checks the grade. If a bottom grade line is used, the man doing the grading can check his own grade. This may also be the case when a top grade line is used and the trench is shallow enough so the grade mark can be seen easily. It is important that material being used as fill have a good moisture content so it will compact well. Do not use dry and dusty material. It will settle when jetting the trench. Before the man in the trench makes his final trim with his shovel, he should pack the material with his feet to make it as firm as possible.

If a pneumatic plate tamper is required, a second grading crew should follow the tamper to make any fills needed for the final trim. If sand or aggregate is required, a hoe or loader should dump material to the man grading. The man on top of the trench shows the operator on the hoe or loader how much to dump and where it is needed.

If you have a choice of the type of aggregate to be used, request ½ inch crushed rock. Fine washed rock or sand could also be used. Crushed rock is an excellent material to use because it makes grading easy, does not settle and compacts with little effort. When the trench bottom is wet or soft, crushed rock must be used to provide a firm bottom for the pipe. The trench should be cut far enough below the final grade to allow for the extra rock needed. If the trench bottom is sponge-like under pressure from your foot, it is too soft for laying pipe.

Most inspectors require that each pipe be checked for grade with a string line. This can also be done with a surveyor's level or a laser beam. Some inspectors will allow the pipe layer to use a carpenter's level to check each pipe if the inspector is confident the trench was graded properly. In this case, the pipe layer can check each pipe with a carpenter's level to be sure it has the correct flow. If you use a carpenter's level, set the first few pipes with the string line so the pipe layer is sure these are at the correct pitch. Check these with the carpenter's level so you know how much bubble tilt is needed. Then set the rest of the line using that same bubble tilt.

The person laying the pipe must have three tools: a square nose shovel, a bar, and a level. When using a bar to shove the pipe together, a block of wood should be placed between the pipe and the bar to prevent the pipe from being chipped. The man helping the pipe layer should have a supply of lubricant and an applicator. If rubber couplings are used, he should have a wrench to tighten the couplings. Figure 24-4 shows the three most common sewer pipe joints. Two men in the trench are all that are needed for laying sewer pipe, with one exception. If rubber couplings are used, the pipe laying may go faster than the pipe layer's helper can tighten the screws. When this happens, a third man must follow to screw the couplings tight. When bell and spigot types are used, the bell always faces upstream so the spigot is pushed into the bell end. If solid coupling type pipes are used, the coupling should be placed on the upstream end of the pipe in the ditch before the next pipe is laid. When using solid coupling or rubber coupling pipe, the pipe usually comes with the coupling already attached to the end. If the first pipe laid has the coupling facing upstream,

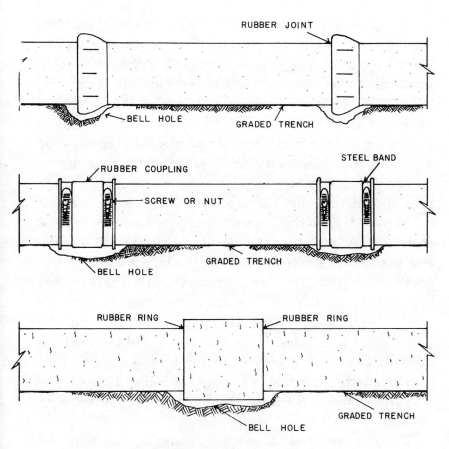

Three types of sewer pipe
Figure 24-4

all couplings will be correct. Before joining bell and spigot or solid coupling pipe together, a lubricant supplied by the pipe manufacturer must be brushed on the rubber end. The bell end is usually the only end that needs lubricating, except when laying solid coupling asbestos pipe. In this case the spigot end must be lubricated. In cold weather, lubricate both ends of the pipe to make joining easier.

The pipe layer must dig a bell hole for each pipe. Bell type pipe needs a larger bell hole than coupling type pipe. See Figure 24-4. The pipe layer and his helper must take great care in keeping the joints straight and free from dirt, sand, or gravel. If any foreign matter gets into the joint, it is likely to leak when tested.

If the rubber ring and spigot end of solid coupling type pipe is not well lubricated or is dirty, the rubber sealing ring may be dislodged when the joint is shoved together. Failure to line up the two ends will also cause the rubber ring to roll out of place. When this happens, the pipe may appear to be seated correctly. But the pipe layer can usually tell that the ring has rolled because he must apply a lot more pressure to get the pipe seated. If more pressure than normal is needed to push the two joints together, the rubber ring has probably rolled from its seat. When this happens, the joints must be pulled apart and cleaned, the ring re-set and the joint re-lubricated before the pipe is shoved together again. It is easy to tell when the rubber coupling or bell and spigot type pipe are not seated correctly. The spigot will not seat fully in the bell.

Every gravity flow sewer line should be laid from the downstream end of the line to the upstream end. It is also advisable to lay through manholes rather than leaving out a section of pipe at each manhole. If there are any sewer laterals to be run from the main, be sure to set the fittings, Y's or Tee's next to the trench where they should be placed. See Figure 24-5. This should prevent the pipe crew from laying straight pipe past the spot where the fitting is to be placed. The detailed drawing on the plans or specifications show the angle the Y's and Tee's must take when laid.

The depth of the trench and the location of parallel pipe

A well prepared set-up for laying sewer pipe. The pipe has been spread correctly and is accessible to the pipe layer and his helper. Notice that the lateral Tee's have been set where they will be laid so they will not be missed. The Tee bell points in the direction it will face when laid
Figure 24-5

determine the angle of rise the service branches will have from the main line to the property line. See Figure 24-6. The specifications tell you how many bends and how sharp a bend can be used on lateral runs. Job specs also show where clean-outs are required. The main must be laid and the Y's

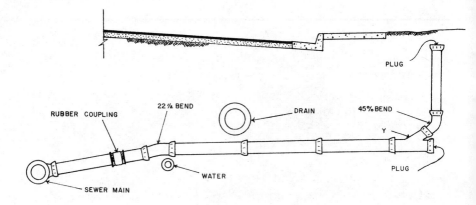

Sewer service
Figure 24-6

and Tee's must be in place before any lateral trenching can begin.

At the end of the day, plug all open pipe ends and place enough dirt or aggregate over the pipe to keep vandals from breaking pipe already laid by dropping pipe or rocks in the trench. If a pipe or fitting in the middle of a line is broken, it can be repaired as follows. Expose and clean the main three joints on each side of the break. Remove all material on top of and on the sides of the main. Cut out the broken pipe and remove the pieces. Five men are needed for inserting a section, two on each side and one in the center of the broken pipe. Lift the sewer line high enough to buckle a new section of pipe in place. This procedure works only on small size pipe that can be lifted by hand. If the pipe is too large for this, proceed as follows. Assuming the pipe is in 6 foot lengths, remove the broken section. Cut the bell off the good end left in the trench. Cut it approximately 1 foot in back of the bell. Now cut 1 foot off the spigot end of the pipe on the upstream side of the broken length. Take a new 6 foot section of pipe that is to replace the broken pipe and add the 1 foot section of bell and the 1 foot section of spigot end to it. Slide two rubber couplings on the two cut ends of pipe. The couplings must be set far enough onto the two ends so that no rubber protrudes past the cut end of the pipe. If the rubber coupling has a rubber strip inside as a divider, cut it out so the whole coupling will slide onto the cut end. The coupling may need lubricating before it will slide fully over the pipe. Lower the pipe with the 1 foot sections and couplings into the space in the main where the broken pipe was removed. Once it is lined up correctly, slide the two rubber couplings over the two standing ends and tighten the couplings to complete the repair. This method works on rigid coupling asbestos cement pipe as well as rubber coupling and bell and spigot pipes.

Placing sewer pipe is a fairly easy operation. If you have problems in joining the pipe or with pipe breakage, call the pipe manufacturer's representative. He can usually give the assistance required to solve the problem.

An efficient sewer laying operation requires planning and coordination. Have the equipment you need and calculate the trench width required. The grade dug by the trencher must be accurate. Don't leave shallow areas that require hand work. The shoring crew must move fast enough to stay ahead of the grading operation. The men grading must have the material they need close at hand. They have to be fast enough to keep ahead of the pipe layers. The pipe layers must have the pipe laid out along the trench correctly so they do not have to wait for pipe. Figure 24-5 shows a well-planned job. The trencher has excellent crumbing shoes to keep the dirt pulled into the bucket line. The man doing the crumbing has no trouble keeping up with the machine. The man behind the bucket line is throwing loose material out of the trench so the bottom will be firm for the crushed rock. One man is checking grade for the man grading. The pipe has been spread correctly with lateral Tee's in place for the pipe layers. The Tee's point in the direction they will be laid.

Once the pipe is laid, it should be partially backfilled. The specifications may call for a sand or aggregate backfill 1 foot over the pipe before native material can be used. If no initial backfill is required, be careful that no soil chunks fall on the pipe. A heavy rock could break the pipe. Shove material into the trench at a 45° angle to the trench. This way the soil will tend to roll over the pipe rather than fall directly on it. Don't cover lateral Tee's if the laterals have not been laid yet. Mark them with a stick or lath and shove a small amount of backfill material over the Tee.

Check the specifications for compaction requirements. If water jetting is allowed, be sure the jet rod is not so long that it will hit and possibly break the pipe.

If the pipe being laid is too large to lay by hand, a crane must be used. Truck mounted cranes are excellent for laying sewer line because pallets of pipe can be carried on the truck bed. The truck then lays each pipe from the supply on the bed.

25

Laying Drain Pipe

As with water and sewer pipe, the first step in laying drain pipe is to prepare for delivery of the pipe to the job site. Be sure the correct equipment is on hand to unload the pipe. To have the correct equipment, you must know the length and weight of the pipe that will be delivered and whether the pipe will be banded on pallets or in single lengths. Plan to use slings or pipe hooks that help avoid damaging the pipe. Many pipe manufacturers will supply the equipment and do the unloading.

Regardless of who unloads the pipe, it is extremely important that the pipe be placed where the laying crew needs it. To find the number of pipe that should be placed between grade stations, divide the length of pipe into the footage between stations. The answer is the number of pipe needed. If the pipe is not distributed correctly, production

A truck mounted crane swings a 42 inch concrete pipe over the trench. Notice that there are four men in the trench and that two men are banding. Look closely and you will see that small piles of cement mortar have been dumped on top of the pipe. These will be used for banding. One of the banders has jumped up on the opposite pipe to help the pipe layers get the pipe turned

Figure 25-1

will be much slower. The distance between the unloaded pipe and the trench is determined by the type of equipment being used to lay the pipe. If you are using a swing crane, the pipe should be placed far enough from the trench to allow driving room between the trench and the pipe. See

Figure 25-1. If a rigid side crane that can only raise and lower is used, the pipe must be unloaded along the trench edge.

Once the pipe has been unloaded, dump any bedding material needed in strategic locations along the trench so it can be dumped or shoveled into the trench with a minimum of effort. The amount of bedding material and initial backfill required should be computed and that amount of material should be dumped between stations or at every other station. If the pipe joints will be the cement type, sand and cement must be dumped in several locations along the trench. Three types of drain pipe are shown in Figure 25-2.

If a bottom grade string is being used, two men can handle the trench grading. If grade is established from a top line over the trench and if the string line is higher than eight feet, have a third man checking grade to speed up the operation. If aggregate is being graded at the trench bottom and if the aggregate is being dumped in the trench with a hoe or loader, the third man can be used in the trench to spread the aggregate ahead of the man grading. One man at the top of the trench can check the grade and direct the hoe or loader.

If excess water is a problem at the bottom of the trench, undercut the final grade enough to get down to firm soil or excavate deep enough so that you can bridge the area with crushed rock. Crushed rock is a good bedding material. It stays firm even when water is running through or over it. Never attempt to lay pipe on a trench bottom that feels spongy under foot.

A well balanced pipe laying operation needs only enough men for shoring, grading, laying, banding, and backfilling pipe to keep up with the trenching operation. There may be

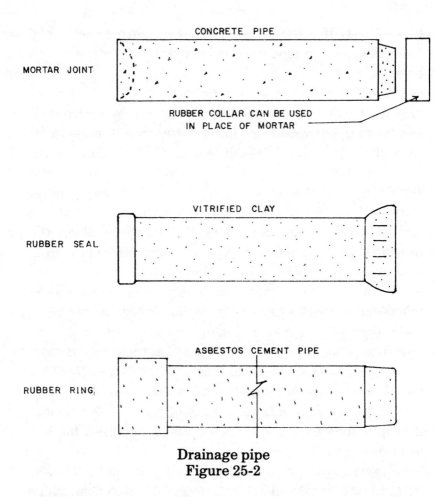

Drainage pipe
Figure 25-2

times when existing utilities hamper the trenching operation to the point where three men can handle the shoring, grading, laying, and banding and still keep up with the trencher.

There are three types of drain pipe in general use and the laying of each varies slightly from the other two.

Slip ring Drain line should always be laid from the downstream end to the upstream end. Be sure the pipe bells are laid in the correct direction. The bells should face upstream. The rubber ring type drain line is joined by lubricating the bell and spigot of the pipe and pushing the two together. A steel bar jabbed in the ground can be used as a lever to force them together. Be sure that no dirt or gravel get in the collar. Each pipe must be lined up straight with the pipe it is joining. Otherwise, joining the two will be difficult if not impossible. If the two pipe are not lined up straight when being joined, the rubber ring may roll from its groove and become crimped. This will result in a leak.

Rubber collar pipe The rubber collar type drain pipe is laid in the same manner as the ring type pipe except that no lubricant is used before joining the pipe. The rubber collar is placed on the pipe in the trench. The pipe being laid is shoved into the collar. Keep the joints free of dirt so they will join completely.

Cement joint pipe Laying pipe that requires a cement joint is much more difficult than laying rubber joining pipe. One man is needed to keep a supply of mortar available for the laying and banding operation. The size of the mortar mixing equipment depends on the size of the pipe and how many feet a day you anticipate laying. For example, if the total production for the day will be about 200 feet and the pipe is 12 inches in diameter or less, use a wheelbarrow for mixing mortar. If 400 feet of pipe will be laid and the pipe is 12 inches in diameter or less, use a mortar box for mixing. If the pipe is 36 inches in diameter and 400 feet will be laid each day, a gas powered mortar mixer should be used.

Laying concrete pipe in a high production operation takes three men in the trench and two above. Two men lay pipe and two men grade the trench. One man out of the trench hooks up the pipe. The second supplies mortar for laying and banding. When 48 inch or larger concrete drain pipe is being laid, two men may be needed for banding, one on each side of the pipe. See Figure 25-1.

Assume that the grading has proceeded far enough down the trench for the pipe laying to begin. The first step in laying cement joint pipe is to mix the mortar. A creamy consistency is needed so the mix will stick to the pipe. The water content is very important. If the mortar is too soft, it will fall off the pipe. If it is too stiff, it will be hard to apply. A good rule of thumb: Take a shovel full of mortar. Hold it three feet above the mortar box and let it slide off the shovel. If it disappears well into the surface of the mortar below, it is about the right consistency. If it stays in a mound on top of the mortar or only partly submerged, it is too dry. If it runs off the shovel, it is too wet. A good sand-to-cement ratio to use is one sack of cement to 20 square nose shovels heaped with sand and one shovel heaped with mortar cream. If plastic cement is used, the mortar cream will not be needed.

The mortar should be passed to the pipe layer and his helper in five gallon buckets. Each bucket should be about ⅔ full. The man or men doing the banding will also need buckets of mortar. The men laying and banding the pipe will also need a half full bucket of water occasionally. The men working with mortar must wear rubber gloves to avoid cement poisoning since they apply the cement to the joints with their hands. The man laying pipe must have a concrete brush nailed to a pole as long as the full length of the pipe. The pipe layer's helper and the man or men banding the pipe need concrete brushes with regular handles. The pipe layer must have a steel bar five or six feet long and 1-1/8 inches in diameter. The pipe layer also needs a shovel.

The first length of pipe is lifted with a cable, preferably

connected to a quick coupler. The hookup man gives hand signals to the crane operator and tells him when to stop lowering the pipe. The first joint of pipe is set into place and unhooked. The grooved or bell end should be facing upstream. The pipe layer digs a small depression for the bell in front of the pipe. The hole should be the width of the portion of pipe that is touching the ground. The pipe layer then puts one or more handfuls of mortar in the small trench that he dug. The amount of mortar depends upon the size of the pipe. He then takes the short handled concrete brush from his helper and splashes water in the pipe groove. He smears mortar in the groove on the bottom half of the pipe. Now the pipe layer moves his buckets and tools back to make way for the second pipe. The pipe layer should check the trench grade where the next pipe will be set to see if it is still level. If the grade has been disturbed from laying the first pipe, he should smooth the trench bottom with a shovel.

By this time the man on top of the trench should have the second pipe hooked up. As the pipe layer steps back, the second pipe is let down into the trench to the point where the pipe layer's helper is standing. Once the pipe being lowered is waist high, the helper motions to stop lowering the pipe. He should then splash water on the spigot end of the pipe and apply mortar to the top half of the spigot. The pipe then is lowered until it is touching the trench bottom lightly. Keep just enough weight on the hoist line so that the pipe can be moved by hand. The pipe layer and his helper guide the spigot end of the second pipe into the groove end of the first pipe. The pipe layer joins the two by pushing the second pipe firmly into the groove of the first pipe with his bar. The two pipes must be lined up straight with each other or the groove and spigot will not slide together easily. If laying pipe on a curve, shove the pipe together first and then turn it to match the radius.

When the joints have gone together correctly, there will be a space between the two pipes: approximately 3/8 inch between pipe up to 12 inches in diameter and 1 to 2 inches between pipe 48 to 60 inches in diameter. If the pipe has been joined correctly, mortar will be squeezed from between the two pipes as they are shoved together. After the pipes are firmly in place, the pipe layer's helper should check the flow of the pipe with a torpedo level. If the flow looks good, he detaches the cable sling from the pipe. The pipe layer then uses his brush on the pole to remove excess mortar at the joint inside the pipe. A sweeping motion completely around the inside of the joint should smooth off any mortar that was squeezed out when the two pipes were joined. Any excess in the pipe should be brushed out. The pipe layer's helper smoothes the mortar that was squeezed from the joint on the outside. The procedure described here should be used each time a new section of pipe is laid.

The man or men banding the joints follow close behind the laying crew. They should splash water on the outside of the joint so it will be damp before applying mortar. Then the banding man takes a handful of mortar in each hand and forces it under the pipe. With the mortar left in his hand, he applies a strip of mortar upward toward the top of the pipe. He uses the other hand full of mortar to do the same thing on the opposite side of the pipe. If the strips of mortar fall short of meeting each other at the top of the pipe, he should apply more mortar using the same procedure until the pipe has been banded all around. Jamming the mortar under the bottom side of the pipe forces the mortar to join with the mortar the pipe layer placed in the small trench while laying the joint. This makes a complete mortar band around the joint. The thickness and width of the band will vary with the pipe size. For example, 12 inch diameter pipe should have a band approximately 4 inches wide and ½ inch thick. A 60 inch diameter pipe needs a band approximately 7 inches wide and 1½ inches thick. After the band has been applied, the person banding should smooth the band with a damp brush. If the pipe laying is proceeding slowly for one reason or another, the banding can be done by the pipe layer's helper after each section is laid. The banding operation should stay a few joints behind the pipe laying so that the banded joints are not jarred as the pipe is pushed together.

The job specifications may call for a cover of some type to protect the bands from drying too quickly. The most common protection used is dirt sprinkled over the band, paper brushed over the band, or sprayed on curing compound.

Once the pipe has been laid, it can be backfilled. The initial backfill must be done cautiously. A sudden surge of material on one side can cause the pipe to roll and crack the bands. If the job specifications require that the backfill be tamped in layers, the specs should also require that the bands have time to cure before any compacting is done. Time for curing is also necessary if water jetting is allowed. If no time is specified, you should wait at least 8 hours before jetting the fill.

If the concrete pipe has to be cut for any reason, a hammer must be used. The size of the hammer required depends on the thickness and size of the pipe. A jackhammer and a chisel may be needed on very large pipe. Be sure that safety glasses are worn when cutting pipe with a hammer. Pipe cutters are available for cutting cast iron, asbestos, and vitrified clay pipe.

When mortar is being used for banding and laying pipe, plan the last batch of mortar mixed so that it is exhausted at the end of the shift. But give the crew enough time for cleaning the mixing equipment, buckets, and gloves before quitting.

26

Constructing Manholes

Excavation for manholes has been explained in an earlier chapter. This chapter explains how to build up the concrete manhole form and set the grade rings that level the manhole cover with the finished grade. Figure 26-1 shows the poured concrete bottom, precast concrete barrels, and cone, and the location of the castings and grade rings.

Manholes that don't have a sump at the bottom should be built to allow a smooth flow of water. This is possible only when the pipe is laid through the manhole bottom. If a side lateral enters the manhole, the lateral should be laid into the manhole to a point where the sweep in the poured bottom will start to enter the main line. See Figure 26-2. If a second lateral enters the manhole from the opposite side, it should be laid the same way. The side lateral pipe can be plugged with a sand bag or anything similar to keep the concrete from running into it when pouring the bottom. Be sure that

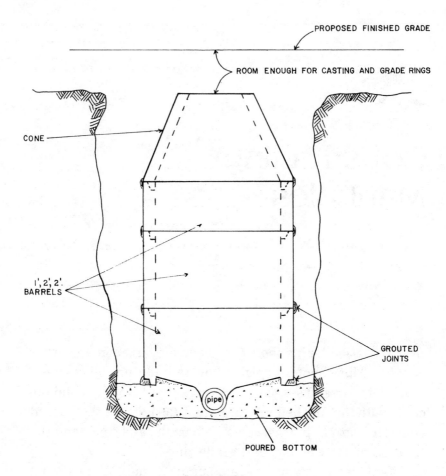

PROPOSED FINISHED GRADE

ROOM ENOUGH FOR CASTING AND GRADE RINGS

CONE

1', 2', 2'
BARRELS

GROUTED
JOINTS

pipe

POURED BOTTOM

**Manhole
Figure 26-1**

the manhole is excavated far enough under the pipe and is the required diameter to comply with the job specifications.

When pouring the manhole bottom, be careful not to pour concrete directly on the pipes that extend across or into the manhole. If the manhole barrels must be set soon after pouring, use 3 percent calcium chloride in the concrete mix

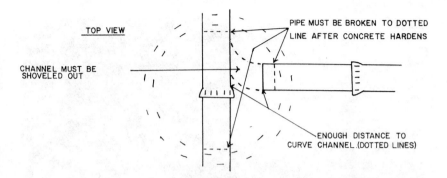

TOP VIEW

CHANNEL MUST BE
SHOVELED OUT

PIPE MUST BE BROKEN TO DOTTED
LINE AFTER CONCRETE HARDENS

ENOUGH DISTANCE TO
CURVE CHANNEL.(DOTTED LINES)

Figure 26-2

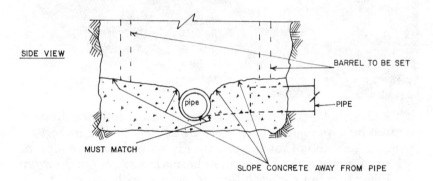

SIDE VIEW

BARREL TO BE SET

PIPE

pipe

MUST MATCH

SLOPE CONCRETE AWAY FROM PIPE

Figure 26-3

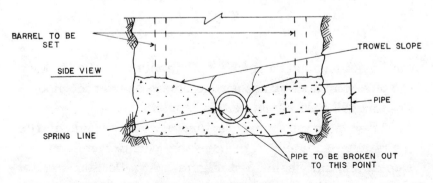

BARREL TO BE
SET

SIDE VIEW

TROWEL SLOPE

PIPE

SPRING LINE

PIPE TO BE BROKEN OUT
TO THIS POINT

Figure 26-4

This is a molded precast bottom. A hand finished bottom
cannot be expected to look quite this smooth. The important
thing to notice is that the channel is as wide as the pipe or
wider at every point. The side channels have a good sweep
to them, turning the side flow to merge with the main line
flow correctly. Notice that the pipe extends into the manhole
bottom only about 4 inches. Leave a hand trowel manhole
bottom looking as much like this as possible
Figure 26-5

for the bottom to get a quick set. Care must be taken to keep
the pipe from floating as the concrete rises under the pipe. A
dry mix will help eliminate this problem.

After pouring concrete just above the top of the
pipe—approximately 2 inches—the channel and bottom
sloping should begin. See Figure 26-3. Do some initial
shaping and then let the concrete dry until it is firm enough

to hold its shape. The final shaping and troweling is then done. It is important to shape the concrete wall at the spring line of the pipe so the bottom has a full channel width as shown in Figure 26-4. If a side lateral is also being channeled in, be sure the concrete is channeled low enough to allow an even flow into the main line.

Do not break out the pipe laid through the manhole until the concrete has hardened—usually not less than 12 hours. If the concrete is not allowed enough curing time before the pipe is broken, the pipe may crack past the poured bottom back into the line. This will cause a leak and the pipe will have to be replaced. After the excess pipe has been broken away, the manhole bottom must be grouted to give the appearance that it has been molded into shape. See Figure 26-5. The tools needed for finishing a manhole bottom are a hammer, a square nose shovel, rubber gloves, and a finisher's trowel or brick layer's trowel. Once the bottom has been troweled as smooth as possible, a fine brush is used to lightly etch the finished surface.

If precast manhole bottoms are available, they should be considered. They are much faster to set. See Figure 26-6. A precast bottom must be placed in an undercut grade. The ground is then brought up to grade with a layer of crushed rock. The pipe is then laid up to the manhole. The precast manhole bottom is lowered into place and connected to the downstream end of the sewer with a short section of pipe with two male ends. Once in place, the upstream pipe is laid into the manhole. A slip ring is used for connecting the male end. When the pipe has been joined to the manhole, the main line can continue. The manhole bottom is finished and ready for setting the barrels. Precast manhole bottoms are ideal when time is important because it eliminates a 2 hour

This precast manhole bottom is a time saver. It should be
used whenever available to speed the completion
of the system
Figure 26-6

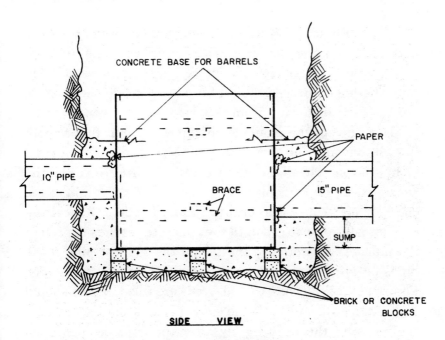

CONCRETE BASE FOR BARRELS

PAPER

10" PIPE

BRACE

15" PIPE

SUMP

BRICK OR CONCRETE BLOCKS

SIDE   VIEW

Form for a manhole with a sump
Figure 26-7

wait between pouring the concrete bottom and setting the barrels.

When specifications require a sump, the pipe should not be laid through the manhole. Instead, pipe extends just far enough into the manhole to reach the inner wall. After the hole has been dug to the depth specified, the bottom is poured. Next, the inner wall forms are inserted and set on the concrete bottom. The inner wall forms can also be raised to the bottom grade as shown in Figure 26-7. The walls and bottom can then be poured at the same time. Wood forms or Sonotube can be used for the inner wall forms. The pipe must be trimmed to length carefully with a hammer so the form has just enough room to slide between the two pipe ends. See Figure 26-7. If there is a space between the form and pipe large enough for the liquid concrete to pass through, use paper from concrete sacks or newspaper to plug the holes. Braces should be placed inside the form to keep it from distorting from the weight of the concrete.

The sides of the manhole should be poured first if both the bottom and the sides are being poured. Pour several inches of concrete all the way around the form. Never let the concrete build up on one side of the form. The weight of the concrete may cause the form to slide off center. Repeat the circular pouring pattern until the concrete reaches the top of the pipe. Always pour to the top of the highest pipe if the pipes entering the manhole are at different elevations. See Figure 26-7. To be sure that all voids are filled, a pole should be used to prod and settle the concrete. Use a hammer to tap the inside of the form lightly after the concrete has been poured. This makes the surface smoother after the forms have been pulled. It is not good practice to use a concrete vibrator for small manhole bottoms when no outside forms are used. A vibrator would cause the concrete to seep into the pipe at the end spaces.

Once the sides have been poured, the bottom can be poured. Be careful when pouring the bottom to keep the concrete just below the bottom form. If the concrete is poured higher than the bottom edge of the form, the form will be very hard to remove once the concrete hardens.

The time needed for curing before the forms are pulled depends on the weather. On a warm day with calcium chloride added to the concrete, three hours is enough if time is important. To test the concrete, bang it with the handle of a shovel. If the handle sinks in or dents the concrete easily, do not pull the forms. If the concrete feels solid, the forms can be removed. Once the forms have been pulled, clean the paper from the ends of the pipe and knock off any rough points of concrete. The manhole can be grouted with a mixture of sand and cement to fill any voids. Finally, brush the concrete smooth.

Now you are ready to set the precast barrels. Check the job specifications for the required joint material. Usually cement or a tar compound is specified. Setting precast barrels is like stacking blocks. Most have a tongue and groove. Place the tongue down and the groove up. The barrels come in various lengths. Choose the correct length for each manhole so that the top barrel will correspond with the finished ground level. See Figure 26-1. If the street or parking area has not been cut to finished grade, have the surveyors set the top of the manhole elevation. If the surveyors are not available, get the elevation from the plans. The plans should give the invert elevation of the pipe at the manhole and also the street or ground elevation. Subtracting the invert elevation from the street elevation will give you the total height of the manhole from the flow line. Now subtract from that figure the thickness of the cast iron manhole casting and the distance from the flow line to the bottom of the first barrel. This gives you the required height of barrels and cones. Any minor adjustment can be made with grade rings 3 or 6 inches high. The job specifications will list a maximum number of grade rings that can be used on each manhole. Be sure the barrels are set high enough to keep the number of grade rings within the maximum set by the specifications.

The generally accepted way of placing manholes in new streets is to leave them deep enough so that they can be paved over. After the paving has been completed, the

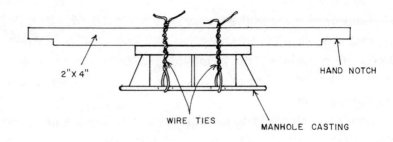

SIDE VIEW

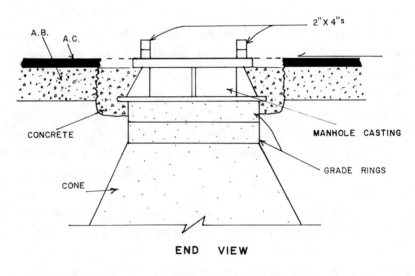

END VIEW

Figure 26-8

manholes are uncovered and the castings are set. To be sure that the castings are level with the pavement, 2 x 4 inch boards are tied with wire to the manhole castings and hung over the grade rings to be poured. See Figure 26-8. When working on a traveled street where the raised casting must be poured and paved around, timing is important. A good four-man crew working an eight hour shift should finish digging, setting, and paving 9 manholes in a day. When paving around manholes, use 1/2 or 3/8 inch aggregate mix and oil the bottom and edges before paving.

27

Pressure Testing Sewer Pipe

There are two ways to test sewer lines for leaks: water tests and air tests. Most specifications require that sewer lines be tested after installation and backfilling are completed. If house laterals are already connected to the new house service lines, they must be plugged before this testing is begun.

To make a water test, the section being tested must be plugged at each end. Both the downstream and upstream manholes must be plugged on the upstream side. See Figure 27-1. Any pipe entering the upstream manhole from the sides must be plugged also. It is good practice to tie a wire or short length of rope to each plug and wrap the wire or rope around a small board. This will keep the plug from being pushed down the pipe by the water pressure if the plug becomes loose. Once the plugs are secure, release water into

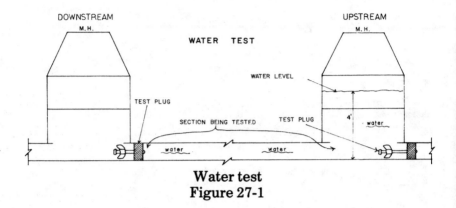

Water test
Figure 27-1

the upstream manhole until the pipe is full and the water rises about four feet up the manhole barrels. Four feet above the pipe is the height required in most specifications. This supplies enough pressure to the main to conduct the test. Four hours should elapse before testing begins so that the pipe and manhole are fully saturated. After four hours, add water to replace water that was absorbed into the pipe and manhole walls. Once the water is again four feet above the pipe, the test may begin.

The amount of leakage allowed will vary from one job specification to the next and depends on the agency involved. The tolerable amount of leakage is usually 500 gallons per mile per day per inch of diameter of the pipe tested. Be sure not to fill the manhole with more water than is required by the specifications. Each extra foot of water increases the pressure in the main substantially. If for some reason a greater head of water is required, the leakage allowed is usually raised 80 gallons for every extra foot raised.

It is to your advantage to test before all the backfill material has been placed, especially if the trench is very deep. The manholes should be built as soon as possible so

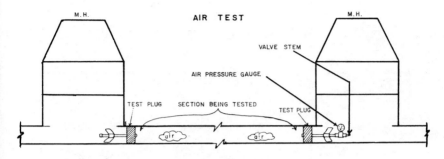

Air test
Figure 27-2

testing can begin. If a leak develops while testing, keep a good head of water in the manhole until the water seeps to the top of the ground. If only a small amount of backfill has been placed, this will occur much sooner and will help you pinpoint the leak faster. If the specifications require that the sewer line be backfilled completely and jetted or compacted before testing, you would be wise to conduct your own test before backfilling and compacting. It is very hard to find a leak during a water test when the trench is deep and has been jetted or compacted.

The air test is the fastest, cleanest, and least expensive way of testing sewer mains and services. The upstream manhole must be plugged on the downstream side. This is the opposite of the water test. The downstream manhole is plugged as in a water test. See Figure 27-2. One of the two plugs set must have a pressure gauge and a valve stem as shown in Figure 27-2. It is not necessary to tie the plugs to a board when air testing. Once the plugs are set, air is pumped into the line. An air nozzle and compressor are needed to fill the line with air. Most agencies require that three to five pounds of pressure be pumped into the main and services. Usually the pressure must be held for three

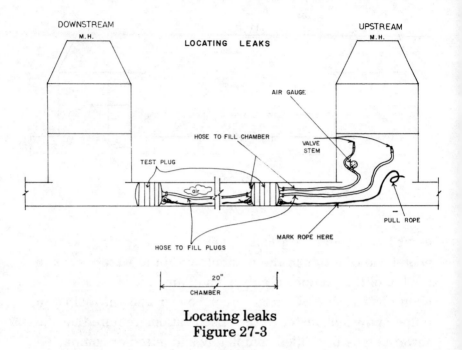

Locating leaks
Figure 27-3

minutes with a maximum pressure loss of 0.5 pounds per square inch. The time the pressure must be held varies with the size of the pipe being tested. The job specifications will list the amount of time and pressure required.

If a leak is detected during the test, it must be located. Even when the main has been jetted or compacted, a leak is easily located if you have the correct apparatus. To locate the leak, remove the two test plugs. A special set of test plugs is used to pinpoint the leak. These special plugs are inflatable and are connected by an air hose and a rope as shown in Figure 27-3. Measure the distance from manhole to manhole and attach enough air hose and rope so that the test plugs can pass through the entire section to the downstream manhole. To get the rope and air lines through the pipe, a small cord is floated down through the line. The end of this line is connected to the plugs and the rope and air lines are pulled back through the line to the downstream manhole. The rope must be pulled until both plugs on the test apparatus enter the pipe at the upstream manhole. One man must be in each manhole until the plugs and line are in place in the main. Once the second plug enters the pipe, locating the leak may begin.

Make a mark on the rope where it enters the pipe at the downstream manhole. Marking the rope is necessary to determine how far the test apparatus has been pulled down the line once a leak is found. The air hose fills both plugs with air to seal off each test section. Usually 20 feet are tested at a time. Once the plugs have been inflated, air is pumped into the second line to pressurize the pipe section between the two plugs. Fill the pipe section between the two plugs to the air pressure specified and wait two minutes. If the pressure holds, release the plug pressure. This also releases the pressure in the pipe. Once the pressure has been released, pull the test plugs ahead 19 feet. Repeat this test cycle until you locate the section that will not hold pressure. Now you have isolated the leak within a 20 foot section. To further pinpoint the leak, move the test plugs ahead three to five feet each time and re-test. Continue this operation until the test pressure holds well again. Now measure the amount of rope that has been pulled from the pipe. If 100 feet has been pulled through, measure 100 feet from the upstream manhole toward the downstream manhole and mark the ground at that point. This will give you the location of the last plug on the test apparatus and will pinpoint the leak within three to five feet—the distance the plugs were moved ahead each time. Continue moving the plugs 19 feet each time until the entire main has been tested for leaks. There may be more than one leak, so don't discontinue testing when the first leak has been located.

After the leak has been pinpointed within three to five feet, check the location of the sewer services. If a service enters the main at that location, the leak could be in the service line rather than the main line. The best procedure to follow in this case would be to expose and check each end of the service line. Check the service where it connects to the main line and at the property line. These are the most likely spots for leaks. If there is not a leak at either end, the entire length of the service line must be checked. Test plugs should be placed at each end of the main and pressure kept in the line. Keep the line under pressure while excavating to make finding the leak easier. The leak may be caused by a stone or dirt wedged in the rubber fitting. In that case, air rushing from the joint will pinpoint the problem. If the leak is caused by foreign matter in a joint, it may be possible to clear off enough room to pull the pipe apart, clean it, and buckle it back together. If a cracked pipe is found, a new pipe can be buckled into place. If the line is too large to buckle a new section in, or if there is not enough room to do so, the cracked section must be cut out and a new section dropped in. When this is done, rubber caulder couplings must be used at each cut end. Correcting leaks is expensive. A little care while laying the pipe will reduce the number of leaks.

28

Drains
And Culverts

Underdrains are designed to collect and carry off water that accumulates under the road surface. This is done by laying a pipe with several holes drilled at the bottom in a shallow trench filled with a permeable material. This is usually rock with a very small amount of sand or fines so that nothing interferes with the seepage of water through the rock. The ditch is filled with rock from 6 inches under the pipe to the top of the ditch. The rock base used on the road should be laid over the ditch rock with no dirt or subgrade material between them. To accomplish this, the subgrade of the road must already have been trimmed and rolled and accepted by the inspector.

The underdrain pipe trench can then be dug. All the dirt from the trench must be hauled off. The trench must be backfilled with rock after the pipe has been laid. The road aggregate is spread when the trench rocking has been

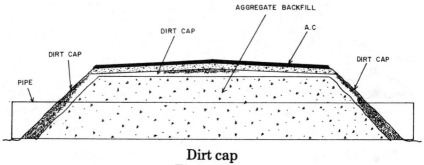

Dirt cap
Figure 28-1

completed. This underdrain causes water running under the base of the road to pass completely through the permeable material. It seeps through holes in the bottom of the pipe and flows through the pipe away from the road base.

Underdrain lines are usually placed at the outside edge of pavement or shoulder. They can be at the high side or low side of the road, depending on the particular seepage problem at that point. Underdrains also are used where there is a subsurface water problem that the engineer is concerned about. In this case, the engineer may design an underdrain with a dirt cap so the road rock and the underdrain rock do not come together. See Figure 28-1. If a dirt cap is called for, the underdrain pipe will be laid after the rough subgrading of the road has been completed. The drain pipe is perforated by drilling and laid with the holes down so the water percolates up through the holes. On some low budget jobs the engineer may eliminate the pipe and use only rock in the ditch as a syphon for the seeping water.

Overside drains are designed to handle water runoff from the road and shoulder surface. The three most common overside drains are corrugated metal pipe, corrugated metal trough, and asphalt pavement. The main point to remember

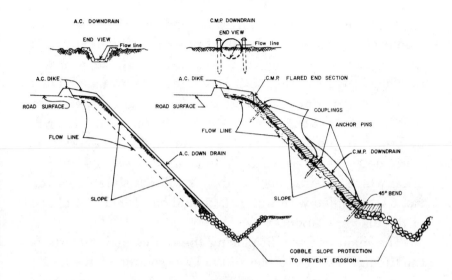

Down drain detail
Figure 28-2

about placing overside drains is that in most cases they cannot be placed until the road is paved and the shoulders have been finished or paved. If they are put in earlier, they will obstruct and slow the work.

If the plans call for an overside drain, a dike will undoubtedly have to be placed. Just before the dike is placed, the overdrains should be put in. If the overside drain is to be an asphalt trough, it can be placed during the dike operation. If the overside drain is a metal pipe, it will be anchored with metal stakes. Overside drain pipe is usually laid at a shallow level with the top of the pipe exposed. Therefore, if the slope is long, it must be braced to keep the pipe from sliding. Figure 28-2 shows an asphalt trough and a metal pipe downdrain.

The overside drain outlet may be designed to run

vertically deep into the fill slope and out the bottom by means of an elbow and length of pipe. In a situation where the overside drain outlet runs deep into the fill slope, it will be placed as the fill slope is built up. It should be capped just below the shoulder grade and uncovered and finished after the shoulders are trimmed. In some cases the drain pipe will be brought up a section at a time during the fill operation.

A metal trough overside drain is primarily a surface type drain which must be anchored so it will not slide down the slope when filled with water. It should be placed just ahead of the dike operation.

The main concern in placing the drains and culverts is the timing. If they are placed too late, you may have to dig through 10 feet of fill to get them in. If they are placed too soon, they may suffer damage from the excavating equipment. The best time to place each culvert depends on many factors. There is no single time during the excavation work that is best for placing culvert. Timing depends on the depth of the culvert. For example, if the fill is to be built up to 60 feet high and the plans show a culvert 5 feet in diameter running under the fill, the most common method would be to build the fill above the culvert to be placed. If the culvert is 5 feet in diameter, build the fill to a point approximately 2 feet above the top of the culvert to be placed, thus giving a seven foot fill. The foreman may elect to build the fill just 4½ feet high to keep from shoring the trench. This would be 6 inches shy of covering the culvert, so when the dirt spread starts again, a dirt ramp will have to be built over the culvert. The ramp should be thick enough so the pipe will not be damaged by the equipment.

On many jobs the drain lines and culverts are not far below subgrade. Therefore the rough grading and the finished subgrade work must be done before the drains can be laid. If the drains are shallow, the material excavated from the ditch must be hauled off or used for road fill. Gravel is used to fill the ditch and backfill the pipe. You can retain enough fill to plug the ends and cap the top as shown in Figure 28-2. In some cases the specifications may call for a shallow pipe to be encased in concrete.

Glossary

Backhoe: Self powered excavation equipment that digs by pulling a boom mounted bucket toward itself. (See picture page 227.)

Balancing Subgrade: Trimming subgrade until there are several areas which are still too high or low, but which when fine trimmed will average out to the finished grade tolerance required.

Bank Plug: Piece of lumber (usually 2'' x 4'') driven into the ground to stand some distance, usually 24'', above ground level. Surveyors place nails in the bank plugs a given distance above the road surface so a string line can be stretched between the plugs to measure grade.

Bench Mark: Point of known elevation from which the surveyors can establish all their grades.

Benching: Making steplike cuts into a slope. Used for erosion control or to tie a new fill into an existing slope.

Bitch Pot: Name used for an oil pot when it contains asphaltic emulsion (oil mixed with water).

Bones: Rocks in the aggregate base which have come to the surface and separated from the finer material. Such a surface is called a "bony" grade.

Boot: A lath set behind the hub by the grade setter when there are obstructions blocking the line of sight to the hub. Grade setter draws a horizontal line on the lath 1 foot or more above the hub and shoots grade from this line.

Boot Truck: Another name for an oil truck with a spray rack for spraying asphalt oil.

Borrow Site: An area from which earth is taken for hauling to a jobsite which is short of earth needed to build an embankment.

Catch Basin: A complete drain box made in various depths and sizes. Water drains into pit, then from it through a pipe connected to the box.

Catch Point: Another word for hinge point.

Center Line: The point on stakes or drawings which indicate the half way point between two sides.

Chip Seal: Process in which fine crushed rock is spread on asphalt oil and then rolled.

Chokers: Road shoulders that are to remain higher than the subgrade level.

Clear and Grub: To remove all vegetation, trees, concrete, or anything that will interfere with construction inside the limits of the project.

Compactor: A machine for compacting soil. Can be pulled or self powered. The latter have wheels to help compaction. (See picture on page 149.)

Crows Foot: A lath set by the grade setter with markings to indicate the final grade at a certain point.

Crumbing Shoe: Metal arm-like attachment on wheel trenchers to keep loose earth at the trench bottom pulled back into the digging buckets.

Curb Shoe: A device bolted to the blade of a grader when grading curbs to help the blade match the shape of the curb bottom.

Dikes: Raised sections built onto the sides of roads to control water runoff and erosion.

Disc: One or more rows of plate-shaped steel wheels, about 3/16" thick, which cut into the earth and roll it over and thus mix in the soil.

Elevation Numbers: The vertical distance above or below sea level.

Embankment: Area being filled with earth.

Feathering: Raking new asphalt to join smoothly with the existing asphalt.

Finished Grade: Any surface which has been cut to or built to the elevation indicated for that point.

Grade: The surface of a road, channel, or natural ground area. Usually means the surface level required by the plans or specifications.

Grade Break: A change in slope from one incline ratio to another.

Grade Lath: A piece of lath that the grade setter has marked to indicate the correct grade to the operators.

Grade Pins: Steel rods driven in the ground at each surveyor's hub between which a string is stretched at the grade indicated on the survey stakes.

Grade Rod: A small length of round or rectangular wood or metal used in place of a ruler for checking grades.

Grader: A power excavating machine with a central blade that can be angled to cast soil to either side. Has an independent hoist control on either side. Also called a "blade."

Guinea: A survey marker driven to grade and colored with blue paint or crayon. Used for finishing and fine trimming.

Guinea Hopper: A member of the grading crew who uncovers the blue topped stakes and signals the blade operator to cut or fill as required.

High Centered: Condition in which the tracks of equipment sink into soft soil. The under carriage resting on soil prevents the equipment from moving out of the soft area.

Hinge Point: A point indicating where the fill slope stops and the road or shoulder grade begins. Sometimes called the "catch point."

Hoe: A track mounted, self powered shoveling machine that digs by pulling a boom mounted bucket toward itself.

Hubs: Point of origin stakes which identify a point on the ground. The top of the hub establishes the point from which soil elevations and distances are computed.

Hyprochlorite Tablets: Tablets placed inside each joint of water pipe for chlorination and purification of water.

Information Stake: Explains in surveyor's code what grades are to be established and the distances to them.

Kicker Blocks: Cement poured behind each bend or angle of water pipe for support. Sometimes called "thrust blocks."

Lane Delineator: A cylinder approximately 3 feet high with a rubber base. Used in a series to control traffic.

Lift: Any layer of material or soil placed upon another.

Lug Down: A slowdown in engine speed due to increasing the load beyond capacity. Usually occurs when heavy machinery is crossing soft or unstable soil or is pushing or pulling beyond its capacity.

Mat: Asphalt as it comes out of a spreader box or paving machine in a smooth, flat form.

Maximum Density and Optimum Moisture: The highest point on the moisture density curve. Considered the best compaction of the soil.

Median: The unpaved section between two or more lanes down the center of a highway.

M.E.E. Pipe: Pipe that has been milled on each end and left rough in the center. M.E.E. stands for "machined each end."

M.O.A. Pipe: Pipe that has been milled end to end. M.O.A. stands for "machined over all" and allows easier joining of the pipe if the length must be cut to fit.

Moisture Density Curve: A graph plotted from tests to determine at what point of added moisture the maximum density will occur.

Natural Ground: The original ground elevation before any excavation has been done.

Nuclear Test: A test to determine soil compaction by sending nuclear impulses into the compacted soil and measuring the returned impulses reflected from the compacted particles.

Oil Pot: Small tank on wheels that can be towed. Has a compressor to supply pressure so road oil can be sprayed from it with a hose and spray nozzle.

Paddle Wheel Scraper: An excavating machine which uses a conveying device to dislodge soil and move it into the bowl.

Pneumatic Tired Roller: A roller with rubber tires commonly used for compacting trimmed subgrade or aggregate base. (See page 115.)

Popcorn: A name given to open graded asphaltic concrete having ¾'' aggregate with very little fine material.

Pug Mill: A rectangular box on wheels containing rows of power driven steel arms that churn dirt and a mixture (usually lime) as it is pulled along the ground.

Pumping: A rolling motion in unstable ground. Usually óccurs when heavy equipment passes over.

Quarter Crown: The area between the center line and the curb or shoulder running parallel to it.

Raveling: A cumulative process in which the rock separates from the finer material on the road surface because of car and truck traffic.

Right Of Way Line: A line on the side of a road marking the limit of the construction area and, usually, the beginning of private property.

Rippers: Teeth-shaped attachments added to digging equipment for digging through hard pan or rocky soil.

RS: Reference stake, from which measurements and grades are established.

Sand Cone Test: A test for determining the compaction level of soil.

Scraper: A digging, hauling, and grading machine having a cutting edge, a carrying bowl, a movable front wall, and a dumping mechanism. (See pages 66, 70.)

Sheepsfoot Roller: A compacting roller with feet expanded at their outer tips, used in compacting soil. (See page 141.)

Spoil Site: Area where unsuitable or excess excavation can be disposed of.

String Line: A nylon line usually strung tightly between supports to indicate both direction and elevation, used in checking grades or deviations in slopes or rises.

Structure Section: Includes all the road material placed from the subgrade level to the finished road surface.

Subgrade: The uppermost level of material placed in embankments or left at cuts in the normal grading of a road bed. This becomes the foundation for aggregate and asphalt pavement.

Summit: The highest point of any area or grade.

Super: A continuous slope in one direction on a road.

Swale: A shallow dip made to allow the passage of water.

Swedes: A method of setting grades at a center point by sighting across the tops of three lath. Two lath are placed at a known correct elevation and the third is adjusted until it is at the correct elevation.

T-Bars: T-shaped bars used in place of steel pins to support a string line over trenches, providing a direct overhead string line to the trenches.

Tangent: A straight line from one point to another which passes over the edge of a curve.

Tied Out: The process of determining the fixed location of existing objects (manholes, meter boxes, etc.) in a street so that they may be uncovered and raised after paving.

Toe of Slope: The bottom of an incline.

Track Loader: A loader on tracks used for filling and loading materials. (See pictures on pages 161, 162.)

"Typical" Drawing: End section view of a street or highway usually showing half of the road if both sides are the same.

Vertical Curve: Indicates a curvature in a horizontal line to a higher or lower elevation.

Vibratory Roller: A self-powered or towed device which mechanically vibrates while it rolls to increase compaction. (See picture on page 148.)

Windrow: The spill-off from the ends of a dozer or grader blade which forms a ridge of loose material. (See picture on page 182.)

Abbreviations

AB	Aggregate Base	ID	Inter Connect
AC	Asphalt Concrete	ID	Inside Diameter
ACP	Asbestos Cement Pipe	IE	Invert Elevation
ASB	Aggregate Sub Base	LBS	Pounds
BC	Begin Curve or Back of Curb	LF	Lineal Foot
BL	Bench Mark	LS	Lump Sum
BSP	Black Steel Pipe	LT	Left
BV	Butterfly Valve	LTB	Lime Treated Base
CB	Catch Basin	MH	Manhole
CF	Cubic Feet	OC	On Center
CIP	Cast Iron Pipe	OD	Outside Diameter
CISP	Cast Iron Soil Pipe	PB	Pull Box
₵	Center Line	PCC	Portland Cement Concrete
CMP	Corrugated Metal Pipe	PG	Projected Grade
CP	Concrete Pipe	PI	Point Indicated or
CTB	Cement Treated Base	V	Angle Point Indicated
CU	Conduit	PL	Property Line
CV	Check Valve	PMP	Perforated Metal Pipe
CY	Cubic Yard	PSI	Pounds Per Square Inch
DBL	Double	PVC	Polyvinyl Chloride Plastic Pipe
DI	Drop Inlet	R=	Radius
DR	Driveway	RCB	Reinforced Concrete Box
EA	Each	RCP	Reinforced Concrete Pipe
EC	End of Curve	RP	Reference Point or Radius Point
EG	Existing Grade	RT	Right
EL	Elevation	R/W	Right-of-Way Line
EMB	Embankment	S=	Slope
EP	Edge of Pavement	SD	Storm Drain
EXC	Excavation	SF	Square Foot
FD	Floor Drain	SG	Subgrade
FG	Finished Grade	SS	Slope Stake or Sewer Service
FL	Flow Line	STA	Station
FH	Fire Hydrant	SY	Square Yards
FS	Finished Surface	TBC	Top Back of Curb
GAL	Gallon	TC	Top of Curb
GB	Grade Break	VB	Valve Box
GD	Gutter Drain	VC	Vertical Curve
GP	Grade Plain	VCP	Vitrified Clay Pipe
GSP	Galvanized Steel Pipe	WSP	Welded Steel Pipe
GV	Gate Valve	WV	Water Valve
HP	Hinge Point	YD	Yard Drain
IC	Inter Connect	yd	Yard

Index

Practical References for Builders

Building Layout

Shows how to use a transit to locate the building on the lot correctly, plan proper grades with minimum excavation, find utility lines and easements, establish correct elevations, lay out accurate foundations and set correct floor heights. Explains planning sewer connections, leveling a foundation out of level, using a story pole and batter boards, working on steep sites, and minimizing excavation costs. **240 pages, 5½ x 8½, $11.75**

Concrete Construction & Estimating

Explains how to estimate the quantity of labor and materials needed, plan the job, erect fiberglass, steel, or prefabricated forms, install shores and scaffolding, handle the concrete into place, set joints, finish and cure the concrete. Every builder who works with concrete should have the reference data, cost estimates, and examples in this practical reference. **571 pages, 5½ x 8½, $17.75**

Rough Carpentry

All rough carpentry is covered in detail: sills, girders, columns, joists, sheathing, ceiling, roof and wall framing, roof trusses, dormers, bay windows, furring and grounds, stairs and insulation. Many of the 24 chapters explain practical code approved methods for saving lumber and time without sacrificing quality. Chapters on columns, headers, rafters, joists and girders show how to use simple engineering principles to select the right lumber dimension for whatever species and grade you are using. **288 pages, 8½ x 11, $14.50**

Construction Estimating Reference Data

Collected in this single volume are the building estimator's 300 most useful estimating reference tables. Labor requirements for nearly every type of construction are included: site work, concrete work, masonry, steel, carpentry, thermal & moisture protection, doors and windows, finishes, mechanical and electrical. Each section explains in detail the work being estimated and gives the appropriate crew size and equipment needed. Many pages of illustrations, estimating pointers and explanations of the work being estimated are also included. This is an essential reference for every professional construction estimator. **368 pages, 11 x 8½, $18.00**

Spec Builder's Guide

Explains how to plan and build a home, control your construction costs, and then sell the house at a price that earns a decent return on the time and money you've invested. Includes professional tips to ensure success as a spec builder: how government statistics help you judge the housing market, cutting costs at every opportunity without sacrificing quality, and taking advantage of construction cycles. Every chapter includes checklists, diagrams, charts, figures, and estimating tables. **448 pages, 8½ x 11, $24.00**

National Construction Estimator

Current building costs in dollars and cents for residential, commercial and industrial construction. Prices for every commonly used building material, and the proper labor cost associated with installation of the material. Everything figured out to give you the "in place" cost in seconds. Many time-saving rules of thumb, waste and coverage factors and estimating tables are included. **512 pages, 8½ x 11, $16.00.**

Process & Industrial Pipe Estimating

A clear, concise guide to estimating costs of fabricating and installing underground and above ground piping. Includes types of pipes and fittings, valves, filters, strainers, and other in-line equipment commonly specified, and their installation methods. Shows how a take-off is consolidated on the estimate form and the bid estimate derived using the complete set of manhour tables provided in this complete manual of pipe estimating. **240 pages, 8½ x 11, $18.25**

Construction Superintending

Explains what the "super" should do during every job phase from taking bids to project completion on both heavy and light construction: excavation, foundations, pilings, steelwork, concrete and masonry, carpentry, plumbing, and electrical. Explains scheduling, preparing estimates, record keeping, dealing with subcontractors, and change orders. Includes the charts, forms, and established guidelines every superintendent needs. **240 pages, 8½ x 11, $22.00**